Forward Deployed Engineer Rule Book

The Future Employment for All Software Engineers

A Lawrence

Forward Deployed Engineer (FDE) Rule Book
The Future Employment for All Software Engineers

This book is a work of nonfiction but reflects the author's opinions and interpretations of industry practice. It is not legal advice, financial advice, or professional consulting guidance. The author and publisher disclaim liability for any actions taken based on this material.

Publisher: A Lawrence
ISBN: 978-1-970328-09-7

United States of America

This edition includes revisions and expansions reflecting updated field practices and training material.

"This book is for educational purposes only. Opinions are the author's own."

Dedication

To every new engineer stepping into the field.

May you have the courage to leave the lab,
the patience to listen before you fix,
and the persistence to carry problems
until they yield.

The world needs your hands, your mind,
and your presence.

Go to where the work is.
Build what endures.

TABLE OF CONTENTS

Preface

"Before the rules, learn the language.
Before the language, learn the field."
— Forward Deployed Engineer Rule Book

This book began with a frustration: too many problems are being solved from too far away. Companies bring in outsiders who diagnose from dashboards and design fixes without ever touching the system itself. The results are predictable — solutions that look clean on paper and fail in reality.

The Forward Deployed Engineer exists because this gap is no longer tolerable. Real problems demand real presence. You cannot know a system until you walk through it. You cannot change it until you've carried part of its weight.

Consider the scale of the opportunity. There are over 430 federal agencies in the U.S., and nearly 91,000 units of government — states, counties, municipalities, school districts, and special-purpose entities. Not one has been systematically penetrated by an FDE. Every one is open terrain for immersion. And beyond government, entire industries — insurance, real estate, oil and gas, healthcare, hospitality, agriculture — remain largely untouched. The field is vast, and the need is urgent.

What follows is not a manifesto or a theory. It is a rule book — forged from practice, fieldwork, and failure. The rules are short, sharp, and meant to be carried with you. They are reminders that the work of an FDE is less about tools or titles and more about posture: going where the problem lives, listening before fixing, and moving forward even when the way is unclear.

This book is written for those who want to do the work right.

Introduction

"In the land of the blind, the one-eyed consultant is crowned king.
But the Forward Deployed Engineer learns to see with both eyes —
by stepping into the field itself."
— Forward Deployed Engineer Rule Book

This is not a book of theory. It is a book of practice.
The Forward Deployed Engineer (FDE) is not a job description you will find in every company. It is a way of working, a stance toward problems, and a discipline for solving them where they actually live.

Most organizations try to solve their toughest challenges from a distance. They hire outside experts, commission endless reports, or build dashboards that reduce complexity to numbers. These efforts often fail because they mistake visibility for understanding. The truth is always messier.

The FDE operates differently. Instead of staying in the lab, they embed themselves in the field. They join the rhythms of the organization, walk the floor, listen to the operators, and see firsthand where systems bend or break. They do not diagnose from afar; they live inside the problem until its shape becomes clear.

This book is written as a rule book, but not in the sense of rigid laws. The rules here are lessons forged in the field — guides to posture, practice, and presence. Each part moves deeper into the craft:

- Part I – Why FDE Matters Now explains why this role exists and why the old models no longer suffice.
- Part II – Anatomy of the Role breaks down - the posture and practices that define an FDE in action.
- Part III – Challenges, Failures, Fixes confronts the realities of systems under stress.
- Part IV – The Future of Engineering Work looks ahead to AI, scale, and global complexity.
- Part V – Appendices & Case Studies provides examples, templates, and lessons drawn from the field itself.

What you will not find here are shortcuts. Being an FDE takes more than cleverness. It requires humility, patience, and the willingness to shoulder problems until you understand them. It is slow work before it is fast work, listening before fixing, and trust before change.

If you are a new engineer wondering how to make an impact, this book will give you a path. If you are a leader wondering why your organization keeps missing the mark, it will show you why.

The Forward Deployed Engineer is not the hero of the story. The field is. This book is about how to stand in it — and how to fix forward.

Glossary

"Before the rules, learn the language.
Before the language, learn the field."
— Forward Deployed Engineer Rule Book

The Forward Deployed Engineer lives in two worlds: the world of people and the world of systems. Each has its own vocabulary. Misunderstand the words, and you misunderstand the work.

This glossary exists to give clarity before the rules. The terms here are not academic. They are field-forged. Each one carries the weight of practice — the language spoken when systems break, when fixes are improvised, and when trust is on the line.

Read them first. Carry them with you.
Because in the field, words are tools.

How to Use This Glossary

This glossary is designed for navigation, not memorization. Terms are written to be read in isolation, revisited as needed, and cross-referenced during real work—not studied front to back. Many concepts reinforce each other; use the "**See also**" references to move laterally between ideas as situations demand. If a term feels abstract, return to it after encountering it in a case, rule, or failure. This glossary reflects field-tested language and judgment, not theory, and is intended to support decision-making under real constraints rather than provide definitive or exhaustive definitions.

■ FORWARD DEPLOYED ENGINEER (FDE) GLOSSARY (with cross-references added)

■ Adaptation

Definition:
The ability to shift with the field instead of resisting it.
Example:
The FDE adapted the workflow overnight when operators changed their process.
Example:
At Devon Energy, hybrid engineers adapted data tools for field operators within weeks instead of years.
See also: Iteration, Fix Forward, Systems Under Stress

■ Adoption Accelerator

Definition:
The FDE as the force that speeds up organizational uptake of technology.
Example:
FDEs trained teams on live projects, making adoption inevitable instead of optional.
See also: Embedded Go-To-Market, Trust, Partnerships

■ AI as Teammate

Definition:
Treating AI as a partner that amplifies direction, not as magic.
Example:
The FDE guided the AI to spot anomalies while people made the final calls.
See also: Human-in-the-Loop, Agentic System, Product Taste

■ AI-Native Product
Definition:
A product designed from the ground up around AI capabilities, not retrofitted automation.
Example:
The team rebuilt workflows instead of automating broken ones.
See also: Agentic System, Integration Depth, Systems of Record

■ Alpha
Definition:
A raw first version tested in the wild to expose hidden flaws.
Example:
The alpha build failed immediately, revealing what the system truly needed.
See also: Failure in the Field, Fix Forward, Iteration

■ Agentic System
Definition:
A system that can plan, execute, adapt, and retry autonomously.
Example:
When the workflow broke, the system replanned instead of halting.
See also: Exception Handling, Systems Under Stress, Fix Forward

■ Apprenticeship
Definition:
The FDE's lifelong practice of learning from the field.
Example:
Even after ten years, the FDE approached deployments as an apprentice.
See also: Immersion, Reading the Field, Secret Degree

■ **Bake-Off**
Definition:
A direct, side-by-side comparison against an incumbent system in real conditions.
Example:
The bake-off exposed failure points hidden by slide decks.
See also: Trust Gap, Pilot Trap, Product Taste

■ **Bridge Role (Glue Role)**
Definition:
The connective role between customer pain and engineering solution.
Example:
The FDE translated operator complaints into system-level fixes.
See also: Product–Problem Translator, Partnerships, Reading the Field

■ **Champion (Internal)**
Definition:
An internal employee who protects and advances adoption.
Example:
The champion ensured the deployment survived early resistance.
See also: Field Champion, Trust, Adoption Accelerator

■ **Challenges**
Definition:
Real-world barriers that test both systems and engineers.
Example:
Regulation, culture, and broken tooling became the true challenges.
See also: Hard Problem, Risk, Systems Under Stress

■ Codegen Tools

Definition:

AI tools that generate, modify, or debug code.

Example:

Codegen tools enabled overnight fixes during live incidents.

See also: Vibe Coding, Prompt Grind, Iteration

■ Corporate FDE

Definition:

An engineer embedded with enterprise clients to accelerate adoption and defend moats.

Example:

Corporate FDEs won deals by solving problems faster than incumbents.

See also: Embedded Go-To-Market, Moat Through Deployment, Trust

■ Debugging Under Fire

Definition:

Fixing live systems under pressure while failures unfold.

Example:

The FDE stabilized the system mid-incident with incomplete data.

See also: Systems Under Stress, Fix Forward, Human Factor

■ Demand Signal

Definition:

Real-world pull that reveals what the field actually needs.

Example:

Shadow spreadsheets revealed the true demand signal.

See also: Pain Point, Reading the Field, Stories as Signals

■ **Do Things That Don't Scale**
Definition:
Manual, high-effort work done early to understand reality before automation.
Example:
The FDE personally onboarded users instead of automating prematurely.
See also: Immersion, Apprenticeship, Embedded Go-To-Market

■ **Embedded Go-To-Market**
Definition:
Adoption driven by integration into daily workflows, not sales theater.
Example:
Adoption succeeded because the product lived inside operations.
See also: Adoption Accelerator, Trust, Integration Depth

■ **Entry Path**
Definition:
How an FDE gains access: hired inside or invited in.
Example:
A licensed professional quietly mapped systems before acting.
See also: Secret Degree, Skin in the Game, Mandated Authority

■ **Eval (Evaluation)**
Definition:
A structured test that measures system correctness under real conditions.
Example:
Evals revealed silent failures hidden by demos.
See also: Prompt Grind, Fix Forward, Systems Under Stress

■ **Exception Handling (Operational)**

Definition:

Defined responses for when automation fails.

Example:

Exception handling prevented system collapse during drift.

See also: Agentic System, Failure in the Field, Risk

■ **Failure in the Field**

Definition:

Breakdowns that reveal truths no design doc can.

Example:

The failure exposed human queues, not technical bugs.

See also: Failure Is Information, Fix Forward, Systems Under Stress

■ **Field Champion**

Definition:

A frontline operator validating real-world usefulness.

Example:

The field champion surfaced issues dashboards missed.

See also: Champion (Internal), Human Factor, Trust

■ **First Deployment Engineer**

Definition:

The original archetype of the FDE embedded from day one.

Example:

They lived onsite, fixing problems overnight.

See also: Immersion, Skin in the Game, Debugging Under Fire

■ Fix Forward

Definition:
Learning from failure and shipping improvements instead of rolling back.
Example:
The team fixed forward and stabilized the system.
See also: Iteration, Failure Is Information, Adaptation

■ Global Engineering

Definition:
Designing systems that survive across cultures and regulations.
Example:
The workflow adapted across five countries.
See also: Scaling, Risk, Systems Under Stress

■ Government FDE

Definition:
An FDE embedded inside public institutions with mandated authority.
Example:
They navigated compliance without halting deployments.
See also: Mandated Authority, Risk, Boundaries

■ Hard Problem

Definition:
A challenge that resists shortcuts and exposes assumptions.
Example:
The system failed because no one understood it end-to-end.
See also: Challenges, Systems Under Stress, Risk

■ **Human Factor**
Definition:
People who keep systems alive when automation fails.
Example:
Handwritten logs saved the deployment.
See also: Trust, Field Champion, Systems Under Stress

■ **Human-in-the-Loop**
Definition:
Humans supervising or intervening in automated decisions.
Example:
Critical decisions remained human-approved.
See also: AI as Teammate, Risk, Trust Gap

■ **Hybrid Engineer**
Definition:
An engineer blending software, systems, and domain expertise.
Example:
Domain knowledge enabled faster fixes than code alone.
See also: Polymath Engineer, Secret Degree, Immersion

■ **Immersion**
Definition:
Embedding until the problem reveals itself.
Example:
The FDE lived the workflow instead of studying reports.
See also: Reading the Field, Apprenticeship, Skin in the Game

■ **Integration Depth**

Definition:

How deeply a system connects into real workflows.

Example:

Shallow integrations failed under pressure.

See also: Systems of Record, Embedded Go-To-Market, Moat Through Deployment

■ **Iteration**

Definition:

Progress through repeated small improvements.

Example:

Each release fixed one real problem.

See also: Fix Forward, Adaptation, Alpha

■ **Mandated Authority**

Definition:

Formal power to enforce adoption.

Example:

Authority opened doors; immersion kept systems alive.

See also: Government FDE, Entry Path, Boundaries

■ **Moat Through Deployment**

Definition:

Defensibility earned by hard deployments, not models.

Example:

Years in the field created an uncopyable moat.

See also: Skin in the Game, Integration Depth, Moat vs. Margin

■ **Pain Point**
Definition:
Lived friction that defines where intervention matters.
Example:
Waiting 30 minutes was the real pain point.
See also: Demand Signal, Human Factor, Risk

■ **Pilot Trap**
Definition:
Confusing pilots with real adoption.
Example:
Paid trials never converted to production.
See also: Embedded Go-To-Market, Trust Gap, Adoption Accelerator

■ **Polymath Engineer**
Definition:
An engineer fluent in tech, domain, and people.
Example:
They navigated policy, code, and users simultaneously.
See also: Hybrid Engineer, Product Taste, Secret Degree

■ **Product Taste**
Definition:
Judgment about what feels trustworthy and usable.
Example:
Better taste beat better models.
See also: Trust, Bake-Off, AI as Teammate

■ **Product–Problem Translator**
Definition:
Turning messy field pain into clear engineering work.
Example:
The FDE reframed complaints into solvable problems.
See also: Bridge Role, Reading the Field, Demand Signal

■ **Prompt Grind**
Definition:
Sustained refinement needed to reach reliability.
Example:
Weeks of grind turned demos into systems.
See also: Eval (Evaluation), Fix Forward, Iteration

■ **Reading the Field**
Definition:
Direct observation instead of dashboards.
Example:
Walking the floor revealed hidden truths.
See also: Immersion, Stories as Signals, Human Factor

■ **Risk**
Definition:
Unavoidable exposure inherent to deployment.
Example:
Naming risk early prevented silent failure.
See also: Systems Under Stress, Boundaries, Failure in the Field

■ **Rule Breaker**

Definition:

Knowing when rules stop serving progress.

Example:

Rigid processes nearly killed the deployment.

See also: Rule Maker, Exception Handling, Fix Forward

■ **Rule Maker**

Definition:

Creating structure that stabilizes systems.

Example:

Clear handoff rules prevented outages.

See also: Rule Template, Boundaries, Trust

■ **Rule Template**

Definition:

Reusable guidance forged through practice.

Example:

"Ship small" became a rule template.

See also: Iteration, Fix Forward, Rule Maker

■ **Scaling**

Definition:

Extending solutions without breaking them.

Example:

What worked locally failed at scale.

See also: Systems Under Stress, Global Engineering, Integration Depth

■ Scaling with Empathy

Definition:

Scaling without abandoning people.

Example:

Training prevented adoption collapse.

See also: Trust, Adoption Accelerator, Human Factor

■ Secret Degree

Definition:

Hidden technical training carried into another profession.

Example:

Credentials granted access; skills revealed systems.

See also: Entry Path, Polymath Engineer, Skin in the Game

■ Skin in the Game

Definition:

Personal stakes that make immersion real.

Example:

Mistakes had consequences.

See also: Immersion, Moat Through Deployment, Risk

■ Speed vs. Precision

Definition:

Balancing momentum with durability.

Example:

Fast enough to help, precise enough to last.

See also: Iteration, Product Taste, Trust

■ **Stories as Signals**
Definition:
Treating stories as operational data.
Example:
A story revealed adoption friction.
See also: Reading the Field, Demand Signal, Stories from the Field

■ **Stories from the Field**
Definition:
Accounts revealing truths dashboards hide.
Example:
A single story reshaped the roadmap.
See also: Stories as Signals, Human Factor, Risk

■ **Systems of Record**
Definition:
Authoritative platforms underlying operations.
Example:
Winning required integration into records.
See also: Integration Depth, AI-Native Product, Trust

■ **Systems Under Stress**
Definition:
The true state of a system revealed by failure.
Example:
Load exposed hidden fragility.
See also: Failure in the Field, Risk, Debugging Under Fire

■ Templates

Definition:

Starting frameworks that must adapt.

Example:

Templates accelerated launch but required change.

See also: Rule Template, Adaptation, Iteration

■ Trust

Definition:

The currency that enables change.

Example:

Trust came from showing up during outages.

See also: Partnerships, Human Factor, Product Taste

■ Trust Gap

Definition:

Hesitation replacing human judgment with automation.

Example:

Comparisons closed the gap.

See also: Human-in-the-Loop, Bake-Off, Trust

■ Uncertainty

Definition:

The constant unknowns of field work.

Example:

The FDE moved forward anyway.

See also: Risk, Systems Under Stress, Adaptation

■ **Vibe Coding**

Definition:

Exploratory building using AI tools.

Example:

Ideas emerged before formalization.

See also: Codegen Tools, Prompt Grind, Iteration

■ **Wrapper Fallacy**

Definition:

Belief that AI systems lack defensibility.

Example:

Deployment difficulty disproved the fallacy.

See also: Moat Through Deployment, Integration Depth, Pilot Trap

MINI INDEX — CROSSWALK OF TERMS

Adaptation
→ Iteration, Fix Forward, Systems Under Stress
Adoption Accelerator
→ Embedded Go-To-Market, Corporate FDE, Trust
AI as Teammate
→ Human-in-the-Loop, Product Taste, Agentic System
Alpha
→ Iteration, Failure in the Field, Systems Under Stress
Apprenticeship
→ Immersion, Entry Path, Secret Degree
Bridge Role (Glue Role)
→ Product–Problem Translator, Partnerships, Reading the Field
Challenges
→ Hard Problem, Risk, Systems Under Stress
Corporate FDE
→ Moat vs. Margin, Embedded Go-To-Market, Entry Path
Debugging Under Fire
→ Systems Under Stress, Fix Forward, Human Factor
Demand Signal
→ Stories as Signals, Pain Point, Reading the Field
Entry Path
→ Secret Degree, Apprenticeship, Skin in the Game
Failure in the Field
→ Fix Forward, Human Factor, Systems Under Stress
First Deployment Engineer
→ Immersion, Debugging Under Fire, Skin in the Game
Fix Forward
→ Iteration, Failure Is Information, Adaptation
Global Engineering
→ Scaling, Risk, Systems Under Stress

MINI INDEX — CROSSWALK OF TERMS (continued)

- **Perfection vs. Progress** → Iteration, Fix Forward
- **Product–Problem Translator** → Partnerships, Bridge Role (Glue Role)
- **Reading the Field** → Immersion, Stories from the Field
- **Risk** → Deployment, Systems Under Stress, Pain Point
- **Rule Breaker** → Rule Maker, Trust
- **Rule Maker** → Rule Breaker, Rule Template
- **Rule Template** → Templates, Scaling
- **Scaling** → Global Engineering, Moat vs. Margin
- **Scaling with Empathy** → Trust, Partnerships, Hybrid Engineer
- **Secret Degree** → Entry Path, Apprenticeship, Skin in the Game
- **Skin in the Game** → Secret Degree, Human Factor, Immersion
- **Speed vs. Precision** → Iteration, Risk
- **Stories as Signals** → Stories from the Field, Demand Signal
- **Stories from the Field** → Case Studies, Failure in the Field, Stories as Signals
- **Systems Under Stress** → Debugging Under Fire, Risk
- **Templates** → Rule Template, Scaling
- **Trust** → Partnerships, Human Factor
- **Uncertainty** → Adaptation, Hard Problem

Part I – Why FDE Matters Now

SECTION A1 — AI REALITY CHECK

Why Forward Deployed Engineers Exist

AI Didn't Fail. Organizations Did.

Every major wave of artificial intelligence adoption has followed the same arc:

1. Excitement
2. Pilots
3. Demos
4. Disappointment

The tools improved. The models scaled. The compute exploded. And yet—real impact stubbornly failed to materialize.

This is not a model problem.

It is an **organizational problem**.

Across industries, most AI initiatives stall not because the technology is insufficient, but because it is **misaligned with reality**: how people actually work, how decisions are actually made, and how systems actually fail.

AI systems are often deployed near the work, but not *inside* it.

That gap is where value goes to die.

The Myth of "Just Add AI"

Many organizations implicitly believe AI adoption works like software installation:

- Select a tool
- Integrate an API
- Train a model
- Roll it out

This mindset treats AI as a **feature** instead of a **capability**.

In practice, AI systems do not succeed by existing. They succeed by being:

- Correctly scoped
- Properly constrained
- Embedded into workflows
- Continuously corrected

None of those are technical steps alone.

They are **field problems**.

Why Pilots Fail at Scale

Most AI pilots succeed in isolation and fail in production for predictable reasons:

- The pilot was built using **idealized data**
- The workflow was **simplified for demo**
- Edge cases were ignored
- Exceptions were handled manually
- Users adapted themselves temporarily

Once deployed, reality returns.

The inputs drift.
The incentives misalign.
The "one weird exception" becomes the norm.

Without someone embedded in the environment—observing, questioning, correcting, and iterating—the system decays rapidly.

This is why organizations report:

- High AI spend
- Low AI return
- Growing internal skepticism

The technology is blamed.

The real issue is **absence of ownership in the field**.

The Deployment Gap

There is a structural gap between:

- **Model builders and model users**
- **Strategy decks** and **operational reality**
- **What leadership thinks happens and what actually happens**

That gap cannot be closed from a dashboard.

It must be closed by a human who:

- Sits with operators
- Watches failures occur
- Traces decisions backward
- Translates ambiguity into structure
- Turns rules into working systems

That role is not accidental.

It emerged because organizations needed it.

Why This Role Didn't Exist Before

Traditional software tolerated abstraction.

AI does not.

Machine learning systems are probabilistic, contextual, and sensitive to small changes in input. They amplify ambiguity instead of hiding it.

This makes **field alignment non-optional**.

Organizations discovered—often painfully—that without someone accountable for the last mile:

- Models were correct but useless
- Predictions were accurate but irrelevant
- Automation increased risk instead of reducing it

The Forward Deployed Engineer exists because **someone has to own reality.**

The Forward Deployed Engineer's Mandate

A Forward Deployed Engineer is not there to:

- Sell AI
- Evangelize AI
- Defend AI

They are there to make it **work**.

That means:

- Understanding the domain before changing it
- Observing workflows before automating them
- Listening before proposing solutions
- Fixing forward instead of redesigning endlessly

This book exists because the industry learned—through failure—that no amount of intelligence compensates for absence in the field.

The world does not wait for clean handoffs. Systems fail in real time, and the cost of delay is measured in broken trust, lost revenue, or human consequences. Traditional engineering roles were built for stability: requirements are gathered, designs are drafted, and solutions are delivered from a safe distance. That model no longer fits.

Today's problems are tangled in the field. They cross organizational lines, involve messy human workflows, and reveal themselves only under pressure. No amount of abstract planning will uncover the truth. You must be present where the work happens.

That is why the Forward Deployed Engineer matters now.

The FDE is not defined by tools, titles, or programming languages. They are defined by posture. They go where the problem lives, embed themselves in the system, and earn the trust of the people who carry its weight every day. They do not observe from the outside — they carry part of the load, so they can learn what truly needs to be fixed.

Companies have tried to outsource this posture to consultants and analytics dashboards. The results are shallow. You cannot understand a system by staring at a chart. You cannot repair a workflow you have never walked. You cannot change what you refuse to live.

The FDE exists because the gap between problems and solutions is too wide to bridge from afar. The role demands engineers who are willing to step into uncertainty, shoulder ambiguity, and work side by side with the field.

This is not a luxury. It is a necessity. Organizations that fail to embed engineers into their hardest problems will keep mistaking symptoms for causes. Those who succeed will not only fix what's broken, they will build systems that endure.

The Forward Deployed Engineer is the answer to a world that no longer waits.

Forward Deployed Engineer vs. Management Consultant

Dimension	Management Consultant	Forward Deployed Engineer
Primary Output	Recommendations	Working systems
Core Deliverable	Decks, memos, frameworks	Code, workflows, artifacts
Definition of "Done"	Strategy approved	System adopted in the field
Timeline Orientation	Engagement-based (weeks/months)	Deployment-based (until it works)
Where Work Happens	Boardrooms, workshops	Live environments, operations
Relationship to Authority	Relies on sponsorship	Operates without hierarchy
Evidence Standard	Consensus, benchmarks	Field behavior, system outcomes
Failure Mode	Good advice, no adoption	Broken system, visible impact
Accountability	Advisory	Operational
After Delivery	Exits with recommendations	Stays through iteration and repair
Response to Resistance	Escalate or reframe	Instrument, observe, adapt
Learning Mechanism	Retrospectives	Live feedback loops
Risk Carried	Reputation	System performance
Primary Optimization	Decision quality	Adoption under real conditions

Chapter 1 – The End of the Old Engineer

"You cannot shape reality from behind a desk.
Step into the field,
or the problem will remain a stranger."
— Forward Deployed Engineer Rule Book

For decades, software engineers lived in silos. Success was measured in commits, tickets closed, and velocity charts. That world is ending.

AI has shattered the distance between software and reality. Models no longer stay in neat boxes; they permeate workflows. Legal contracts, medical charts, balance sheets, and maintenance logs are all potential training data. Every department is suddenly a site for AI deployment.

This creates a new risk: the distant engineer becomes the distant AI engineer. Without immersion, biases go undiscovered, trust erodes, and costs multiply. The FDE's immersion is the essential last mile of AI risk mitigation.

AI requires immersion. It requires trust. It requires translation. The old engineer wrote code. The new engineer must build bridges.

Myth vs. Reality: The Value of CS/CE Degrees

Myth: "Computer Science and Computer Engineering degrees are dead."
Reality: They are more valuable than ever — but only when paired with immersion.
A CS or CE degree alone no longer guarantees impact. But as a Secret Degree, it becomes a hidden lens. It lets you see patterns, workflows, and pain points that others miss.

- In finance, a CPA with a CS degree can map inefficiencies invisible to auditors.
- In medicine, an MD with data science skills can redesign diagnostics workflows.
- In trades, a licensed electrician with coding fluency can automate safety checks.

The degree isn't dead. It's transformed. Alone, it risks irrelevance. Combined with immersion, it becomes the most powerful credential in the field.

Chapter 2 – The Rise of the Forward Deployed Engineer

"A tool can be borrowed.
A mindset must be forged.
Keep sharpening the one thing no one can take from you."
— Forward Deployed Engineer Rule Book

The first experiments in this model came from Palantir. Their early-2000s intelligence software was too abstract to succeed on its own. Engineers had to embed inside agencies, shadowing analysts, interrogators, and case officers. They learned by watching and built by improvising.

Palantir called these people Forward Deployed Engineers—half engineer, half anthropologist. Embedded in the field, they prototyped side by side with users. Adoption soared where pure product teams had failed.

The model spread. OpenAI, Scale AI, Ramp, and Baseten all hired “FDEs” to accelerate adoption. The pitch was simple: you can’t scale AI by throwing code over the wall. You need engineers in the field.

But execution fell short. Many FDEs became little more than consultants with laptops—embedded, yes, but powerless. Observers, not officers. Without mandate, legitimacy, or authority, they lacked teeth.

The idea was right. The structure was incomplete.

Chapter 3 – FDE as a Mandated Officer, Not Consultant

"FDE is not a title on a badge.
It is a way of standing—
in the middle of the problem,
with both feet on the ground."
— Forward Deployed Engineer Rule Book

This Rule Book argues for the next evolution: **the institutional FDE**.

The true FDE cannot be an external consultant parachuted in by procurement. They must be an **internal appointment**, chartered by the Board, appointed by the CEO, or in government, by Congress. Their legitimacy must flow from the top.

They must also be **dual-credentialed**: software engineers who are also licensed insiders. A JD + CS in law, an MD + DS in medicine, a CPA + SE in finance, a licensed tradesperson + CS in HVAC or aviation. Without this dual identity, they cannot build credibility.

Above all, they must carry **authority to act**. Observation and advice are not enough. The FDE must have enforcement power: the ability to demand access, override obstruction, and The Forward Deployed Engineer must have explicit, documented authority within their assigned scope to remove blockers. Without this, the role collapses into consultancy.

This makes the FDE not just a new job description, but a new class of officer inside organizations. Just as the CFO institutionalized financial oversight and the CTO institutionalized technology strategy, the FDE must be institutionalized as the **bridge officer of AI adoption**.

Palantir stumbled onto the prototype. The future demands the full model.

MARKET VALIDATION

Why the Forward Deployed Engineer Role Is Exploding

Demand Is Not Theoretical Anymore

For years, the Forward Deployed Engineer role existed quietly inside a small number of organizations. It was rarely named, often misunderstood, and almost never standardized.

That has changed.

Over the last several years, postings for roles that match the Forward Deployed Engineer profile have grown **dramatically**—far outpacing traditional software engineering and data science roles. In some segments, postings increased by **several hundred percent** in under a year.

This growth is not driven by hype.
It is driven by **failed deployments**.

Organizations are discovering—often after expensive missteps—that hiring more model builders does not fix adoption, and hiring more consultants does not fix execution.

They need engineers who can operate **inside the problem**.

From "Data Science" to "Deployment Science"

Early AI hiring focused on abstraction:

- Better models
- Better benchmarks
- Better architectures

That phase is largely complete.

Today's bottleneck is not intelligence.
It is **integration**.

As a result, organizations are shifting away from roles that optimize models in isolation and toward roles that:

- Sit with users
- Translate needs into constraints
- Adapt systems to reality
- Own outcomes beyond demos

The market is not abandoning technical skill.
It is demanding **applied technical judgment**.

That judgment only emerges through field exposure.

Why Traditional Roles Don't Scale Here

Many organizations attempt to solve this gap using existing roles:

- Product managers lack technical depth
- Consultants lack ownership
- ML engineers lack domain immersion
- Platform teams lack context
- Leadership lacks time

Each group touches part of the problem.
None own the whole thing.

Forward Deployed Engineers exist because **no existing role was accountable for the entire lifecycle:**
from vague problem → deployed system → operational stability.

The market is selecting for roles that close that loop.

Why This Growth Is Structural, Not Cyclical

This is not a hiring trend that will reverse when budgets tighten. In fact, the opposite is true.

When organizations face pressure:

- They cannot afford failed pilots
- They cannot tolerate brittle automation
- They cannot absorb reputational or operational risk

Under constraint, they prefer:

- Fewer tools
- Fewer abstractions
- Fewer intermediaries
- More accountability

Forward Deployed Engineers reduce risk by **owning ambiguity early**, when it is still cheap.

That makes the role more valuable—not less—during downturns.

Why New AI Platforms Increase FDE Demand

Ironically, better AI platforms increase the need for Forward Deployed Engineers.

As tools become easier to use:

- More people attempt deployment
- More non-obvious failures emerge
- More edge cases surface
- More organizational friction is exposed

Powerful platforms magnify mistakes as easily as successes.

Someone must be present to:

- Detect misalignment early
- Intervene before scale
- Adapt workflows incrementally
- Prevent "silent failure" modes

That work cannot be centralized.

It must be **forward deployed.**

What the Market Is Actually Hiring For

Despite varied titles, successful postings share the same signals:

- Strong technical foundation
- Comfort with ambiguity
- Willingness to embed with users
- Ability to translate across functions
- Accountability beyond delivery
- Ownership beyond handoff

These are not junior traits.
They are not checklist skills.

They are **field-earned capabilities.**

Organizations are no longer guessing which skills matter.
They are selecting based on hard lessons learned.

The Implication for This Book

The Forward Deployed Engineer is no longer an edge case.

It is a **response to structural failure modes** in modern software and AI deployment.

The next chapter does not argue that the role exists.
It explains **why it was inevitable**.

GOVERNMENT & OCCUPATION SCALE

Where Forward Deployed Engineers Actually Deploy

The Mistake of Thinking Too Small

When people imagine where Forward Deployed Engineers work, they often picture a narrow slice of the economy:

- High-growth startups
- Elite technology companies
- Defense contractors
- Silicon Valley firms

That picture is incomplete.

The largest opportunity for Forward Deployed Engineers is not in software-first organizations.

It is in **process-heavy, regulation-bound, workflow-dense environments** that predate modern software by decades.

The Hidden Scale of Government and Public Systems

In the United States alone, there are **tens of thousands** of distinct governmental entities:

- Federal agencies
- State departments
- Counties
- Cities
- Towns
- Special districts
- Authorities
- Boards
- Commissions

Each operates:

- With legacy systems
- Under legal constraints
- With fragmented data
- With human-driven workflows
- With real-world consequences

These systems are not small.
They are **structural**.

And most were never designed for automation.

Why These Environments Demand FDEs

Government and regulated sectors share key characteristics:

- Ambiguous ownership
- Conflicting incentives
- Manual exception handling
- Incomplete documentation
- High cost of failure
- Slow feedback loops

These are environments where:

- Centralized solutions fail
- "Best practices" don't translate
- One-size-fits-all tools break immediately

They require engineers who can:

- Observe before optimizing
- Respect constraints without being paralyzed by them
- Translate policy into executable logic
- Work patiently inside imperfect systems

That is Forward Deployed Engineering.

Occupational Density and Process Depth

Beyond government, many of the largest deployable domains are occupationally dense:

- Healthcare operations
- Utilities
- Transportation
- Logistics
- Construction
- Compliance
- Inspections
- Licensing
- Benefits administration

These fields share a pattern:

- Thousands of micro-decisions per day
- Heavy reliance on tacit knowledge
- Informal workarounds
- Paper trails that hide real logic

Automation here does not fail because it is hard.

It fails because the work is not visible unless someone embeds.

Why "Digital Transformation" Misses the Mark

Many organizations attempt modernization through:

- Procurement
- Platform rollouts
- Transformation initiatives
- External vendors

These efforts often stall because:

- They abstract too early
- They optimize the wrong layer
- They ignore edge cases
- They fail to earn trust

Forward Deployed Engineers operate differently.

They don't start with systems.
They start with **people**.

They sit next to the work and ask:

- "What actually happens when this breaks?"
- "Who fixes it?"
- "How often does that occur?"
- "What happens if we automate the happy path?"

Only after that do they touch technology.

Why This Matters for Career Strategy

This scale changes the career calculus.

There are far more complex systems in:

- Governments
- Utilities
- Healthcare
- Infrastructure

than in consumer apps.

Yet these environments are **underserved**, under-automated, and desperate for competent field operators.

Forward Deployed Engineers who understand this can:

- Build deep expertise
- Create durable impact
- Avoid crowded labor markets
- Operate where competition is low and stakes are high

This is not a niche.

It is a **map of where the work actually is**.

The Transition Ahead

If Forward Deployed Engineers can operate across thousands of organizations and occupations, a new question emerges:

What is the correct relationship between the engineer and the organization?
Are they inside?
Are they outside?
Are they temporary?
Are they permanent?

That distinction matters more than it appears.

Part II – Anatomy of the Role

Chapter 4 – The Dual-Credential Requirement

"You cannot fix what you refuse to live.
Go where the work is hardest.
See what the slides never show."
— Forward Deployed Engineer Rule Book

When Palantir first embedded engineers into intelligence agencies, one truth emerged quickly: brilliant coders without domain expertise were ignored. Analysts spoke in acronyms, lawyers in statutes, doctors in codes, and tradesmen in tolerances. The engineer sat quietly, lost.

The future FDE cannot afford to be lost. They must be dual-credentialed.

- In law: J.D. + CS
- In medicine: M.D. + DS
- In finance: CPA or CFA + CS
- In trades: licensed plumber/HVAC/electrician + ML/SE

This dual identity is the foundation of trust. Lawyers listen to lawyers. Surgeons respect surgeons. Foremen nod to someone who has held the wrench. Without the credential, the FDE is dismissed as an outsider.

The limitation is a strength. A JD+CS FDE cannot lead in a refinery; an HVAC+CS FDE cannot argue in court. Each FDE is a mission specialist. Precision builds credibility. Depth beats generality.

Chapter 5 – The Language of Trust

"Haste breaks trust.
Hesitation wastes time.
The craft is choosing the cut
that delivers both speed and truth."
— Forward Deployed Engineer Rule Book

Credentials open the door. **Language keeps it open**.

Every profession speaks its own vernacular. Misuse a single term and credibility evaporates. A **'margin' means profit** in finance, whitespace in design, and error in statistics. Confuse them, and you prove you don't belong.

Fluency signals three things:

- **Precision**: knowing what words mean in context.
- **Credibility**: sounding like an insider.
- **Cultural trust**: speaking the values of the profession.

An FDE must listen before coding. Read journals, shadow practitioners, absorb protocols. Translate across silos: law to finance, operations to compliance. Only then will they be trusted to reshape systems.

Chapter 6 – Authority to Act

"Every system whispers its secrets.
Only those who walk among it
can hear the language it speaks."
— Forward Deployed Engineer Rule Book

Early FDEs had empathy, prototypes, and relationships—but no power. They could recommend, but not enforce. Without teeth, they were consultants in disguise.

The true FDE must have **authority to act**:

- **Direct reporting**: to CEO or Congress—no middle managers.
- **Mandated cooperation**: departments are required to comply.
- **At-will enforcement**: ability to remove blockers who sabotage.
- **Budgetary power**: outcome-linked budgets they can direct.

This is not authoritarian—it is clarity. Everyone knows the FDE is an a Mandated Officer, not an advisor. Authority doesn't make the job easy. It makes it possible.

Chapter 7 – Reporting Lines and Legitimacy

"Code that breathes in the field
must endure heat, noise, and chaos.
Write for the storm, not the silence."
— Forward Deployed Engineer Rule Book

Bury an FDE inside IT or HR and they die in that silo. Every department will try to capture them.

The only viable reporting line: **directly to the CEO, with Board ratification**. In government: **Congressional appointment**.

Legitimacy comes from mandate. Without it, the FDE is a tourist. With it, they are a Mandated Officer —free to roam across silos and assemble the enterprise puzzle.

Chapter 8 – The CFDE: A New C-Suite Role

"You are not the savior parachuting in.
You are the bridge,
built plank by plank with trust."
— Forward Deployed Engineer Rule Book

Every transformation produces a new executive seat: CFO, CTO, CRO. AI adoption will require the **CFDE—Chief Forward Deployed Engineer**.

The CFDE is not a glorified engineer. They are an Executive Mandated Officer responsible for ensuring AI adoption across every department. Their powers:

- **Authority**: mandate from the top.
- **Breadth**: deployment across silos.
- **Integration**: harmonizing departmental FDEs.

In the future, no serious enterprise will operate without a CFDE.

Chapter 9 – The FDE Corps: Specialists in Every Department

"Every release is a roll of dice.
The wise engineer knows
what can be lost before the throw."
— Forward Deployed Engineer Rule Book

The CFDE commands a Corps of FDEs, each dual-credentialed in their field:

- **Legal FDE**: JD + CS, trusted in contracts and compliance.
- **Finance FDE**: CPA/CFA + CS, trusted in audits and forecasting.
- **Medical FDE:** MD + DS, trusted in diagnostics and safety.
- **Technical Trade FDE**: Licensed HVAC/plumber/aviation + CS, trusted in infrastructure.
- **HR FDE**: Org Psych + SE, trusted in workforce planning.

Each speaks their domain language. Each carries authority. Together, they form the operational arm of adoption.

When NOT to Use a Forward Deployed Engineer

Knowing the Limits of the Role Is Part of Using It Correctly

The Forward Deployed Engineer Is Not a Universal Fix

Forward Deployed Engineers are effective because they operate inside ambiguity, surface uncomfortable truths, and turn reality into executable systems.

That does **not** make them appropriate for every situation.

In fact, many of the most damaging FDE failures occur not because the engineer performed poorly—but because the role was deployed into an environment that could not support it.

This section exists to prevent that mistake.

An FDE Cannot Replace Missing Leadership

If an organization lacks:

- A clear decision owner
- Willingness to make tradeoffs
- Authority to change workflows

A Forward Deployed Engineer will become a **buffer**, not a solution.

In these environments, the FDE absorbs tension while real decisions are deferred indefinitely.

Do not deploy an FDE when leadership expects the engineer to compensate for indecision.

An FDE Is Not a Firefighter for Structural Problems

Forward Deployed Engineers are often requested during moments of crisis:

- Failed launches
- Regulatory pressure
- Public scrutiny
- Operational breakdowns

If the expectation is:

- "Fix this quietly"
- "Make the problem go away"
- "Stabilize without changing anything"

The role is being misused.

FDEs surface reality.
They do not conceal it.

An FDE Cannot Create Trust Where It Is Punished

Forward Deployed Engineers depend on truthful information.

If the environment:

- Punishes bad news
- Penalizes uncertainty
- Rewards narrative alignment over accuracy
- Discourages admitting failure

Then elicitation will fail—no matter how skilled the engineer.

Do not deploy an FDE into a culture that treats truth as a liability.

An FDE Is Not a Substitute for Product Ownership

Some organizations attempt to use FDEs to fill gaps in:

- Product management
- Strategy
- Roadmapping
- Prioritization

This is a category error.

FDEs:

- Execute against constraints
- Surface misalignment
- Translate reality into systems

They do not decide *what the organization wants to be.*

When product ownership is absent, the FDE will drift into scope definition by default—and fail under contradictory expectations.

An FDE Should Not Be Used for "Instant ROI" Theater

If success is defined as:

- A demo by a fixed date
- A visible prototype regardless of readiness
- Short-term optics over long-term viability

Then the incentives are inverted.

Forward Deployed Engineers require permission to:

- Delay scaling
- Expose edge cases
- Pause execution
- Walk away from bad paths

Without that permission, the role collapses into **consulting with code**.

An FDE Cannot Operate Without Boundaries

Do not deploy an FDE when:

- Access rules are undefined
- Data ownership is unclear
- Ethical constraints are negotiable
- "We'll sort it out later" is the norm

Forward Deployed Engineers must be able to explain—after the fact—why every access decision made sense.

If that explanation would be uncomfortable, the environment is not ready.

The Quiet Anti-Pattern: Using an FDE to Delay Accountability

One of the most subtle misuses of the role occurs when organizations say:
"*Let's put an FDE on it and see what they find.*"
...and then do nothing with the findings.

In these cases, the FDE becomes:

- A diagnostic without treatment
- A truth generator without authority
- A witness without recourse

This is not deployment.
It is deferral.

The Test Before Deployment

Before assigning a Forward Deployed Engineer, ask:

- Who owns the outcome after handoff?
- What decisions is the FDE allowed to trigger?
- What happens if the truth is inconvenient?
- What changes are explicitly in-bounds?
- What happens if the best answer is "don't build this"?

If these questions do not have answers, the role should not be deployed yet.

A Professional Boundary, Not a Rejection

Saying "this is not an FDE-appropriate situation" is not a failure.

It is evidence of maturity.

Forward Deployed Engineers create the most value when they are deployed deliberately—into environments that can absorb truth, act on it, and sustain change.

Anything else risks burning out the engineer and leaving the organization unchanged.

Transition Back to the Book

When the environment is ready, the next question becomes critical:

Should the Forward Deployed Engineer sit **inside** the organization—or remain **outside** it?

That distinction shapes incentives, authority, and long-term impact.

Organization Readiness Scorecard for Forward Deployed Engineering

Assessing Whether the Environment Can Support the Role

Why This Scorecard Exists

Forward Deployed Engineers do not fail in isolation.

They fail when the **environment rejects reality faster than the engineer can surface it.**

This scorecard allows organizations to assess—honestly—whether deploying an FDE will create value or friction.

It is not a hiring tool.

It is a **deployment readiness check**.

How to Use This Scorecard

- Score each dimension from **0 to 2**
- Be conservative
- Any score of **0** is a warning
- A total score below **10** indicates high risk

This assessment should be completed before assigning an FDE—not after problems appear.

Executive Sponsorship

Question:
Is there a named leader who owns outcomes, not just interest?

- **2** — Clear sponsor with decision authority and follow-through
- **1** — Supportive but indirect or shared ownership
- **0** — No clear owner; decisions float

Risk if missing:
Findings go nowhere. The FDE becomes advisory only.

Authority to Change Workflows

Question:
Can the FDE trigger real operational change?

- **2** — Explicit permission to modify workflows and systems
- **1** — Changes allowed but politically constrained
- **0** — Observation only; execution discouraged

Risk if missing:
Truth surfaces, but nothing moves.

Tolerance for Inconvenient Truth

Question:

How does the organization respond to bad news?

- **2** — Failure is surfaced and addressed
- **1** — Failure acknowledged but softened
- **0** — Failure punished or ignored

Risk if missing:

Elicitation collapses. Signal degrades.

Access Boundaries Are Defined

Question:
Are data access rules explicit and documented?

- **2** — Clear access rules, escalation paths, and ownership
- **1** — Informal norms but no documentation
- **0** — “Figure it out as you go”

Risk if missing:
Ethical drift, trust erosion, and audit exposure.

Ownership After Handoff

Question:

Who owns the system after the FDE rotates out?

- **2** — Named owner with time and authority
- **1** — Ownership assigned but under-resourced
- **0** — No clear owner

Risk if missing:

Systems decay or quietly fail.

Willingness to Pause or Exit

Question:
Can the organization accept "don't scale this"?

- **2** — Explicit permission to stop or pivot
- **1** — Reluctant acceptance
- **0** — Scaling expected regardless of evidence

Risk if missing:
Sunk-cost escalation and brittle deployments.

Documentation Is Valued

Question:

Is structured documentation expected and used?

- **2** — IRAC-style reasoning welcomed and referenced
- **1** — Documentation exists but rarely revisited
- **0** — Knowledge lives in meetings and memory

Risk if missing:

No continuity. No learning.

Field Access Is Real

Question:
Can the FDE observe the work directly?

- **2** — Direct access to operators and artifacts
- **1** — Filtered access through intermediaries
- **0** — Access limited to leadership narratives

Risk if missing:
The FDE works on a model of reality, not reality.

Scoring Interpretation

- **14–16** → Strong readiness
- **10–13** → Conditional readiness (deploy carefully)
- **<10** → High risk; address gaps first

A low score does not mean the organization is failing.
It means the role is being deployed too early.

What This Scorecard Is Not

- Not a performance review
- Not a maturity model
- Not a justification document
- Not a guarantee of success

It is a **boundary-setting tool**.

The Professional Use of This Scorecard

Forward Deployed Engineers should feel empowered to say:

"Based on this assessment, deploying me now will likely fail—and here's why."

Organizations that listen will save time, money, and trust.

Organizations that don't will learn the hard way.

Transition Back to the Book

Once readiness is established, a final structural decision remains:

Should the Forward Deployed Engineer be **inside** the organization—or **outside** it?

That distinction shapes everything that follows.

Chapter 10 – The Orchestra Model

"Perfection is paralysis.
Iteration is discipline.
Build, test, and build again."
— Forward Deployed Engineer Rule Book

Left alone, each FDE optimizes for their silo: Legal for compliance, Finance for risk, Medical for safety, Trades for uptime, HR for fairness. All valid—but often in conflict.

The CFDE acts as Bridge Role:

- **The Score**: shared organizational priorities.
- **The Baton**: authority to resolve conflicts.
- **The Harmony**: cross-silo feedback loops.

The result: not soloists, but a symphony. Specialists for depth. CFDE for breadth. Mandate for legitimacy.

Chapter 11 – The "Secret Degree"

Your Computer Science or Software Engineering degree is your hidden card — *the secret degree.* You don't advertise it. You don't lead with it. But it changes how you see every workflow, every bottleneck, every failure. Where others see routine, you see systems. Where others accept friction, you spot patterns.

Only you know the truth:
☛ You are there to study the company or trade as a system, piece by piece, from the inside out.
Strength: unmatched depth. You see politics, pain points, and culture that outsiders miss.
Risk: assimilation. If you forget your secret mission, you may settle into the dysfunction you hoped to fix.
This path only works if you choose a field you actually care about. Otherwise, the grind will break you. Immersion requires skin in the game — your paycheck, your license, your reputation.
Example: A young graduate earns a CPA, joins a Big Four accounting firm, and spends years auditing ledgers. Outwardly, he is just another accountant. Inwardly, he is mapping inefficiencies and planning future fixes. His **SE** degree stays hidden. When he leaves to found a fintech platform, he isn't guessing — he is rebuilding from memory.
This is the long way. But it is the deep way.

Path Two: Invited In (External Expertise)
The sharp road.

Here, you don't apply through HR. You are brought in because you or your company claims an expertise. You are the **outside FDE team**: coders, troubleshooters, translators, sent to diagnose and fix.

- Mindset: surgeon. Cut fast, stitch clean.
- Strength: clarity. You see dysfunction insiders no longer notice.
- Risk: rejection. You may be viewed as a foreign body, tolerated only if you heal faster than you disrupt.

External FDEs thrive when they deliver outcomes immediately. Every fix is a test. Every hour without progress burns credibility. Example: Palantir's early deployments with the U.S. Army. Their engineers were outsiders — coders in combat tents. But by patching systems mid-exercise and living on base, they turned suspicion into trust. They weren't soldiers, but they were present where the failure hurt.

The Gray Zone: Hybrid Entry

Most FDEs eventually drift into the gray zone.

- **An outsider FDE** becomes part of the night crew after months on-site.
- **An insider hire** gets reassigned into a field project and starts seeing with outsider clarity.

Hybrids are powerful because they combine the two. They are trusted like insiders but still ask forbidden questions like outsiders. They balance belonging with perspective.

Entry as a Strategic Choice

The entry path is not random. It reflects **strategic intent**.

- Organizations choose: build FDEs internally, or import them externally.

- Individuals choose: take the long road of hired-inside immersion, or the fast road of external deployment.

The hired-inside route is slower — but deeper.
You will know the company better than its executives, because you lived its workflows. You will see the system from the inside out, and when you leave to build your startup, you will not be guessing. You will be reconstructing from memory.

Bob McGrew, an early Palantir executive, learned this lesson the hard way. The first system he built for a U.S. agency was, in his words, "100% wrong." Why? Because he didn't yet understand the system. FDEs who walk the hired-inside path pay in years, but avoid that fatal mistake.

The Rule of Entry

The path shapes the work.

- As an insider, your power is trust earned slowly.
- As an outsider, your power is clarity and urgency.
- As a hybrid, your power is balance: belonging without blindness.

Whatever the path, one truth remains:
☛ **Before you can fix forward, you must first enter.**

Lessons for FDEs

1. **Credentials are your disguise**. To enter, you must hold the field's own qualifications. Your CS/SE degree stays secret.
2. **Skin in the game matters**. Choose a field you care about. Without interest, immersion collapses.
3. **Clarity is your weapon**. Outsiders cut fast. Insiders must fight to keep perspective.

- Immersion is the moat. The deeper you live inside a system, the harder it is for others to replicate your insight.
- Remember your mission. An insider who forgets they are an FDE becomes just another employee.

Why It Matters

Every story in this book begins with entry:

- Palantir's Army teams were **invited in** as external experts.
- GitLab FDEs were **hired inside** as staff, shadowing workflows to drive adoption.
- Devon Energy's hybrid engineers blurred the line, half domain experts, half coders.
- Immigration deployments demanded persistence: no one asked for help, entry had to be earned.

The way you enter is not logistics. It is destiny.
The path is slow or sharp, long or short, trusted or distrusted.
But without entry, there is no immersion. Without immersion, there is no adoption. And without adoption, there is no FDE.

The entry path is the first rule:
To fix a system, you must first walk through its doors.

📌 Secret Degree — Day-in-the-Life Sidebar

(What the "*Secret Degree*" looks like on a random Tuesday)

8:47 AM — Intake

A regulator flags a contract workflow as "non-compliant." The language is vague. No specific clause is cited.

- Domain move: Read the regulation and identify the operative phrases that trigger enforcement.
- Engineering move: Extract those clauses into a structured checklist that can be applied at scale.

9:30 AM — Scope Check

The organization believes the issue affects "a handful" of contracts.

- Domain move: Ask what legally counts as "material exposure."
- Engineering move: Write a quick script to scan all active contracts for the relevant terms.

The answer is not "a handful."

11:05 AM — Field Reality

Operations insist the workflow has "always worked this way."

- Domain move: Sit with operators and watch the approval process end-to-end.
- Engineering move: Map the actual workflow, not the documented one.

Three manual steps are missing from the system entirely.

12:40 PM — Lunch Is Canceled

A meeting turns adversarial.

- Domain move: Reframe the conversation around obligation, not blame.
- Engineering move: Pull concrete examples that show where the system diverges from the rule.

Arguments stop when evidence appears.

2:15 PM — Translation

Leadership wants a recommendation.

- Domain move: Clarify which risks are legal, which are operational, and which are reputational.
- Engineering move: Model each risk as a discrete failure mode with a corresponding fix.

Not all risks require code.

3:50 PM — Prototype

A permanent fix will take weeks.

- Domain move: Identify what is legally acceptable today.
- Engineering move: Ship a temporary validation check that blocks the worst cases without halting the business.

This is not the final system. It is a boundary.

5:10 PM — Documentation

Decisions are made under uncertainty.

- Domain move: Record what was known at the time, not what is known now.
- Engineering move: Capture the decision in an IRAC+ memo so reasoning survives handoff.

No slide deck is created.

6:02 PM — Exit

Nothing is "finished."

- Domain move: Assign ownership for the next decision point.
- Engineering move: Leave behind artifacts that allow the system to keep learning without you.

The work continues after you leave the room.

Sidebar Close (one line)

The Secret Degree is not about credentials.
It is about operating fluently where rules, systems, and reality collide.

THE FDE EXECUTION LOOP

From Vague Problem to Working System

Why Most Execution Models Break Down

Many organizations believe execution follows a clean sequence:

Requirements → Build → Deploy → Done

Forward Deployed Engineers know this is fiction.

In the field, execution is messy because:

- Problems are ill-defined
- Constraints surface late
- Incentives conflict
- Data lies
- Workarounds exist
- Reality changes midstream

The FDE Execution Loop exists to handle **imperfect information without freezing.**

The FDE Execution Loop (Overview)

Forward Deployed Engineers operate in a loop, not a line:

1. **Scoping**
2. **Proof of Concept**
3. **Production Hardening**
4. **Field Feedback**
5. **Iteration or Exit**

Ownership does not end at deployment.
It ends when the system survives reality.

1. Scoping: Define the Real Problem

Scoping is not about features.

It is about:

- Identifying the decision that matters
- Defining what "better" actually means
- Establishing constraints explicitly
- Mapping who is affected and how

This phase is driven by:

- Elicitation
- Interviews
- Observation
- IRAC+ documentation

A poorly scoped problem guarantees wasted effort—no matter how good the implementation.

2. **Proof of Concept: Test Reality, Not Demos**

A Proof of Concept exists to answer **one question**:

Does this work under real conditions?

FDE PoCs are:

- Narrow
- Fragile
- Honest
- Disposable

They deliberately:

- Use real data
- Include ugly edge cases
- Expose failure modes
- Avoid polish

A PoC that cannot fail is not a PoC.

3. Production Hardening: Pay the Cost Up Front

Most systems fail in production because hardening is postponed.

Forward Deployed Engineers harden early by addressing:

- Data quality
- Latency
- Reliability
- Security
- Monitoring
- Human fallback paths

This phase is unglamorous—but decisive.

Hardening is where:

- Automation becomes trustworthy
- Risk becomes visible
- Ownership becomes real

4. **Field Feedback: Reality Pushes Back**

Once deployed, the field responds.

This response reveals:

- New edge cases
- Misaligned incentives
- Unexpected behavior
- Human adaptations
- Organizational friction

Forward Deployed Engineers stay close during this phase.

They watch:

- What users bypass
- What they complain about
- What they quietly stop using

Feedback is not a meeting.

It is observation.

5. **Iterate or Exit: Decide Deliberately**

Not all systems should scale.

Forward Deployed Engineers decide:

- Iterate → if value compounds
- Exit → if constraints dominate
- Redesign → if assumptions failed

The key is **intentionality**.

Walking away from a bad path early is success—not failure.

Why This Loop Works

The FDE Execution Loop:

- Absorbs uncertainty
- Limits blast radius
- Preserves trust
- Encourages learning
- Prevents sunk-cost escalation

It replaces heroics with discipline.

The Hidden Responsibility

Forward Deployed Engineers are often the only ones who:

- See the full lifecycle
- Experience the pain directly
- Understand tradeoffs holistically

That visibility creates responsibility.

The execution loop ensures that responsibility is exercised—not avoided.

Transition Forward

Execution does not end with one system.

The next challenge is scaling—without losing field alignment.

That requires restraint as much as ambition.

FDE Toolkit Essentials

(Categories, not brands)

Forward Deployed Engineers do not carry a fixed stack.
They carry a **set of capabilities** that let them move between field reality and working systems without friction.

What matters is not what you use, but why you reach for it.

1. **Data Pull & Reality Capture**
Used when: Someone says "we think," "usually," or "in most cases."

- Structured queries against production data
- Exports from systems of record
- Logs, event streams, and audit trails

Purpose: Replace anecdote with evidence before decisions harden.

2. **Rapid Analysis & Sense-Making**
Used when: The problem is visible but not yet shaped.

- Lightweight scripts or notebooks
- One-off transforms and sanity checks
- Quick aggregations that surface outliers

Purpose: See the shape of failure without building infrastructure.

3. **Workflow Glue**

Used when: Systems technically work, but people work around them.

- Small command-line utilities
- Automation wrappers
- Temporary connectors between mismatched tools

Purpose: Bridge gaps long enough to learn where the real seams are.

4. **Field Observation & Capture**

Used when: Documentation disagrees with reality.

- Live observation of workflows
- Notes taken during actual operations
- Screenshots, timestamps, and hand-drawn diagrams (sanitized)

Purpose: Record what actually happens, not what is claimed to happen.

5. **Prototyping & Boundary Tests**

Used when: You need to change behavior before building the final system.

- Minimal internal tools
- Validation checks
- Guardrails that block the worst failures

Purpose: Create boundaries, not perfection.

6. **Documentation as Control Surface**

Used when: Decisions are being made under uncertainty.

- IRAC+ decision records
- Explicit ownership assignments
- Clearly stated unknowns and next evidence

Purpose: Preserve reasoning so learning survives people, pressure, and time.

7. **Observability & Feedback Loops**

Used when: A fix is live, but trust is fragile.

- Monitoring tied to real outcomes
- Signals that operators actually notice
- Feedback channels that bypass hierarchy

Purpose: Detect drift before it becomes doctrine.

8. **Exit Artifacts**

Used when: You are about to leave the system.

- Runbooks that reflect reality
- Decision trails, not just outcomes
- Clear handoff points

Purpose: Ensure the system can keep learning without you.

What This Is Not

- Not a vendor list
- Not a "modern stack"
- Not a career signal

The FDE toolkit is intentionally small, disposable, and replaceable.

Rule

If a tool survives longer than the problem it was built to understand,
it has already become part of the system — and must be treated as such.

MEMORY IS NOT A DATABASE

Why Requirements Fail Even When Everyone Is "Sure"

The Most Common Hidden Failure Mode

Many failed systems share a deceptively simple origin:

The engineer trusted what people remembered.

This is not negligence.
It is a misunderstanding of how human memory works.

Memory is not a recording.
It is a reconstruction.

Every recall is influenced by:

- Context
- Emotion
- Incentives
- Recent events
- The way the question is asked

Forward Deployed Engineers must treat memory as **unreliable input**, not ground truth.

Why Confidence Is a Poor Signal

People are often most confident about:

- Routine tasks
- Long-standing processes
- "The way we've always done it"

Ironically, these are the areas most likely to be:

- Partially automated
- Informally modified
- Quietly bypassed
- Inconsistently applied

Confidence correlates poorly with accuracy.

This is why:

- Two people describe the same process differently
- Stakeholders agree—until artifacts appear
- Documentation contradicts daily practice

None of this implies dishonesty.

It implies **normal cognition under load**.

How Engineers Accidentally Poison Memory

Forward Deployed Engineers unintentionally distort memory when they:

- Ask leading questions
- Summarize too early
- Confirm assumptions out loud
- React visibly to answers
- Signal preferred outcomes

Small cues matter.

When an engineer says,
"*So the system usually works, right*?"
the answer is already compromised.

Narrative First, Structure Later

Human memory retrieves information best as stories, not lists.

That is why the FDE Interview Loop begins with:

- Open narratives
- Uninterrupted walkthroughs
- Concrete examples

Only after the narrative stabilizes does structure emerge.

Attempting to impose structure too early causes:

- Omitted steps
- Collapsed timelines
- Missing exceptions
- False coherence

Forward Deployed Engineers must resist the urge to "clean up" reality prematurely.

Why Artifacts Matter More Than Recollection

Artifacts do not remember.
They record.

Screens, logs, forms, tickets, and reports often contradict verbal descriptions—not because people are lying, but because systems behave differently than expected.

Forward Deployed Engineers privilege artifacts because:

- They capture edge cases
- They reveal timing
- They expose silent failures
- They show workarounds

When memory and artifacts disagree, the artifact becomes the starting point for inquiry.

Practical Implications for FDEs

The Deeper Insight

Memory is not the enemy.

Unexamined memory is.

Forward Deployed Engineers do not distrust people.

They design conversations that compensate for human limits.

That is how signal survives contact with reality

Transition Forward

Even when memory is handled correctly, one failure mode remains especially dangerous:

The engineer **leads** instead of listens.

That mistake deserves explicit prohibition.

APPRENTICESHIP TIME HORIZONS

Why Mastery Takes Years, Not Weeks

The Dangerous Fantasy of Instant Understanding

Modern engineering culture quietly promotes a myth:

If you are smart enough, you can understand any domain quickly.

This belief is comforting—and wrong.

In real organizations, especially those with long histories and real-world consequences, understanding does not arrive through documentation, onboarding decks, or stakeholder briefings.

It arrives through **time**.

Forward Deployed Engineers must reject the fantasy of instant mastery, because acting on partial understanding is often worse than doing nothing at all

Why Complex Domains Behave Like Apprenticeships

Long before software existed, humans solved complex problems through **apprenticeship**.

Not because knowledge was secret—but because:

- Critical details were tacit
- Judgment developed slowly
- Context mattered
- Mistakes were expensive

That pattern has not disappeared.

Modern domains such as healthcare operations, immigration systems,
utilities, logistics, and public administration still function this way.

They cannot be "learned" in weeks.

They must be **absorbed**.

What Apprenticeship Actually Looks Like

True apprenticeship has recognizable features:

- Extended observation before intervention
- Repetition across varied cases
- Exposure to edge cases
- Supervised responsibility
- Gradual autonomy
- Accountability for outcomes

These features exist whether or not an organization names them.

Forward Deployed Engineers enter domains where apprenticeship already exists—they simply step into it intentionally.

Why Shortcuts Fail in the Field

Organizations often pressure engineers to move faster:

- "You've seen enough."
- "Just automate the obvious parts."
- "We'll fix edge cases later."

These shortcuts consistently fail because:

- The "obvious" path is rarely the dominant one
- Exceptions are not rare—they are structural
- Informal workarounds encode real rules
- Trust is lost faster than it is gained

Forward Deployed Engineers learn that **speed without understanding creates rework**.

Patience is not delay.
It is risk reduction.

The Cost of Misaligned Time Horizons

When leadership expects results in weeks while reality requires months or years, predictable failures occur:

- Brittle systems
- Quiet reversions to manual work
- Shadow processes
- Blame shifting
- Loss of credibility

Forward Deployed Engineers serve as a stabilizing force by **resetting expectations**.

They make the invisible visible:

- What cannot be automated yet
- What requires policy decisions
- What depends on trust
- What must be learned slowly

This honesty protects both the engineer and the organization.

Why FDEs Are Not "Slow"

Patience is often mistaken for slowness.

In practice, Forward Deployed Engineers:

- Avoid false starts
- Prevent expensive rollbacks
- Reduce long-term maintenance
- Build systems that survive contact with reality

They may move more slowly at the beginning.

They move **much faster over time**.

That is the difference between building something that demos and building something that lasts.

The Uncomfortable Truth

There is no shortcut to deep domain understanding.

No framework replaces presence.

No model replaces judgment.

No abstraction replaces accountability.

Forward Deployed Engineers accept this reality—and plan accordingly.

This is why they resemble apprentices early on, and trusted operators later.

Transition to Part III

Once an engineer accepts long time horizons, a new challenge emerges:

What happens when things still go wrong?

Because they will.

Part III examines the failures, frictions, and constraints that Forward Deployed Engineers must face—and how to fix forward without losing trust

Part III – Challenges, Failures, Fixes

Chapter 12 – Reading the Workplace: The Hidden Training of a Forward Deployed Engineer

"Before you can fix the system, you must first learn to see it."
— Forward Deployed Engineer Rule Book

The hardest skill an FDE must learn is not coding or design—it's perception. How do you know what's broken? How do you see what others miss?

1. Learning to See

There is no academy for this craft. No bootcamp can teach you how to walk into a foreign workplace and spot the fracture lines. The FDE learns by immersion and by questions—using the classic **Five Ws** and a **How** as a compass:

- Who feels the pain?
- What breaks most often?
- When does friction appear?
- Where does it originate?
- Why has it never been fixed?
- How do people work around it today?

Every employee carries part of the answer. The FDE's first fieldwork is conversation—asking frontline workers to describe their pain points and how they would improve them. Patterns emerge. Pain repeats. The field reveals its map.

Every employee carries part of the answer. The FDE's first fieldwork is conversation—asking frontline workers to describe their pain points and how they would improve them. Patterns emerge. Pain repeats. The field reveals its map.

2. The Field Interview

Treat every interaction as data. Ask without judgment. Record without bias. Your role is not to critique—it is to observe, translate, and later connect what no single person can see alone. Like law enforcement's evidence logs, your notes must be precise, factual, and structured for others to retrace.

3. The Feedback Loop Upward

When reporting to leadership—be it a Board, a CEO, or Congress—the FDE must not present final truth, but open discourse. Encourage decision-makers to critique the analysis publicly, not behind closed doors. Invite counter-proposals. A system improves fastest when daylight reaches every layer.

4. Offering Options, Not Edicts

A true FDE never delivers a single 'answer.' Instead, present multiple pathways—pilotable options that allow comparison, iteration, and eventual adoption. The goal is not consensus—it is experimentation in the open.

5. Continuity and Custody

If an FDE leaves mid-deployment, the mission must not die. Field notes are the connective tissue of trust. They must be:

• Exacting: enough detail for a successor to resume without re-immersion.

• Secure: classified or proprietary data stored under correct authority.

• Ethical: anything suggesting criminal or civil violation must follow mandated reporting lines. Silence creates liability.

The FDE's notebooks are both record and inheritance—the next engineer's map of the system.

6. Toward a Training Standard

Just as FBI or ICE agents study rules of evidence, criminal procedure & investigative techniques, the FDE needs structured training in observation, documentation, and system ethics. This book is only the starting rule set. The next industry frontier will be the creation of FDE academies—institutions that teach the art of immersion, the law of transparency, and the ethics of system stewardship.

Rule Template: Ask the Five Ws and a How Before You Code

Never diagnose from a dashboard. Walk the floor. Ask the **Five Ws** and a **How**. Let the field teach you what the reports forgot.

RULE EXPANSION

Do Not Lead the Witness

This Rule Exists to Protect Reality

Forward Deployed Engineers are hired to observe reality—not to shape it.
Leading the witness is the fastest way to destroy signal while believing you are being efficient.

This rule exists because even well-intentioned engineers accidentally **manufacture agreement**, then build systems on top of it.

Those systems almost always fail.

What "Leading" Actually Means in Engineering

Leading does not require manipulation or bad faith.

Engineers lead the witness when they:

- Propose solutions before understanding the problem
- Frame questions around preferred outcomes
- Use technical language stakeholders mirror without understanding
- Summarize prematurely and ask for confirmation
- Treat agreement as validation

Common examples:

- "Wouldn't it be easier if the system just did this automatically?"
- "So the main issue is performance, right?"
- "This usually works, correct?"
- "We can probably ignore that edge case, yeah?"

Each question nudges the answer.

Why Leading Feels Productive (and Isn't)

Leading produces:

- Faster meetings
- Cleaner notes
- Apparent consensus
- Reduced discomfort

It also produces:

- Missing constraints
- Suppressed exceptions
- False confidence
- Fragile automation

Leading optimizes for **comfort**, not correctness.

Forward Deployed Engineers must optimize for correctness.

The Mirror Effect

Humans are cooperative by default.

When an engineer:

- Signals confidence
- Uses authoritative language
- Suggests direction

Stakeholders respond by:

- Agreeing
- Simplifying
- Avoiding disagreement
- Protecting face

This is not deception.

It is normal social behavior.

The engineer's job is to **design around it**, not blame it.

How Leading Corrupts Requirements

Once a requirement is shaped by leading:

- Stakeholders defend it
- Contradictions disappear
- Edge cases go unspoken
- Revisions feel like reversals

By the time failure appears in production, it is too late to recover the original truth.

The opportunity window has closed.

What FDEs Do Instead

Forward Deployed Engineers replace leading with:

- Open narratives
- Neutral prompts
- Silence
- Clarification questions
- Artifact verification

They ask:

- "What happened next?"
- "Can you show me?"
- "How often does that occur?"
- "What happens when this goes wrong?"
- "Who handles that exception?"

They let reality reveal itself—even when it is messy.

This Rule Is Non-Negotiable

Leading the witness breaks:

- Interviews
- Documentation
- Trust
- Deployments

It also breaks **IRAC+,** because the issue itself becomes contaminated.

Forward Deployed Engineers are expected to:

- Resist premature solutions
- Tolerate ambiguity
- Delay certainty
- Preserve raw signal

This rule is the guardrail that makes everything else in this book possible.

Enforcement in Practice

If you catch yourself:

- Finishing sentences
- Pitching ideas mid-interview
- Explaining the system back to the user
- Asking yes/no questions too early

Stop.
Reset.
Return to observation.

The Cost of Violation

Violating this rule does not fail immediately.

It fails **quietly**, later, when:

- The system doesn't generalize
- Edge cases overwhelm automation
- Users bypass the tool
- Trust evaporates

By then, no one remembers the original conversation.

The damage is already done.

Transition Forward

Even when engineers avoid leading, another trap remains:

Mistaking confidence, fluency, or charisma for truth.

That error deserves explicit treatment.

IRAC+

The FDE Field Documentation Standard

Why Forward Deployed Engineers Need a Documentation Standard

Forward Deployed Engineers do not fail because they lack intelligence.

They fail because **insight evaporates**.

Conversations happen.
Observations are made.
Contradictions surface.
Workarounds are revealed.

And then:

- The engineer rotates out
- The project pauses
- The system changes hands
- The context is lost

What remains is code without memory.

Forward Deployed Engineering requires a documentation standard that preserves thinking, not just artifacts.

That standard is **IRAC+.**

IRAC Is Not Legal Writing. It Is Structured Reasoning.

IRAC—Issue, Rule, Application, Conclusion—did not originate as an academic exercise.

It emerged as a way to:

- Reason under ambiguity
- Separate facts from assumptions
- Make decisions traceable
- Allow others to audit thinking

Those properties map directly to field engineering.

However, classical IRAC is missing one critical element for deployment work:
Context.

SIDEBAR

Behavioral Cues ≠ Truth

Confidence Is Not Evidence

Forward Deployed Engineers frequently encounter people who appear:

- Confident
- Fluent
- Certain
- Calm
- Authoritative

These traits feel reassuring.

They are also **poor indicators of truth.**

Behavioral cues—tone, posture, speed of response, certainty—correlate weakly with accuracy and strongly with familiarity, power, and comfort.

Mistaking confidence for correctness is one of the most common—and costly—errors in field work.

Why Humans Misread Signals

Humans are wired to infer truth from behavior because:

- It worked in small groups
- It reduced social friction
- It conserved cognitive effort

In complex organizations, those instincts fail.

People who speak confidently may:

- Be repeating inherited explanations
- Be shielding uncertainty
- Be incentivized to appear competent
- Be unaware of downstream failures

People who hesitate may:

- Be more accurate
- Be navigating nuance
- Be aware of exceptions
- Be protecting others

Forward Deployed Engineers must override instinct.

Fluency Masks Fragility

Systems that fail most spectacularly are often explained most smoothly.

Fluency hides:

- Unhandled edge cases
- Informal workarounds
- Dependency chains
- Manual interventions
- Institutional memory

Smooth explanations feel complete—even when they are not.

This is why FDEs ask for:

- Walkthroughs
- Examples
- Artifacts
- Failure stories

Not summaries.

Authority Distorts Signal

People with authority often:

- Speak last
- Frame the problem
- Set acceptable narratives

Their perspective matters—but it is incomplete.

Forward Deployed Engineers treat authority as:

- A source of constraints
- Not a source of ground truth

Truth lives closer to the work.

What FDEs Trust Instead

Forward Deployed Engineers privilege:

- Repeated patterns
- Concrete examples
- Artifact consistency
- Cross-account validation
- Failure frequency
- Exception handling

They trust **systems under stress**, not stories under confidence.

Practical Guardrails

When evaluating information:

- Do not rank answers by confidence
- Do not reward fluency
- Do not punish uncertainty
- Do not assume silence equals agreement

Instead:

- Ask for evidence
- Ask for exceptions
- Ask for last failures
- Ask who fixes things when they break

Rule Template: Do Not Confuse Tenure with Signal

Experience can dull perception. People closest to the work—but newest to the system—often see fractures others have learned to ignore.

The FDE does not privilege seniority, confidence, or polish when gathering signal.

Insight frequently comes from those with fresh eyes, incomplete context, or unrefined language.

Insight often arrives unpolished, sometimes from the newest hire. The FDE learns to listen for it anyway.

When evaluating information in the field:
Do not:

- Rank answers by confidence or seniority
- Reward fluency or polish
- Punish uncertainty
- Assume silence equals agreement
- Dismiss input because it comes from a new hire

Instead:

- Ask for evidence
- Ask for exceptions
- Ask about the last failure
- Ask who fixes things when they break
- Ask who noticed the problem first
- Talk to everyone

The goal is not consensus or authority. The goal is truth under real conditions.

Why This Sidebar Exists

This reminder exists to protect FDEs from:

- Charismatic misinformation
- Over-trusting experts
- Under-listening to operators
- Building systems on narratives instead of reality

Behavior is data.

It is not truth.

The Missing Piece: Context

In the field, problems are inseparable from:

- Who you spoke to
- Where the issue occurred
- When it surfaced
- Under what constraints
- Inside which system

Without that information:

- Issues cannot be re-evaluated
- Decisions cannot be challenged
- Fixes cannot be safely revisited
- Handoffs fail

Forward Deployed Engineers therefore use **IRAC+.**

The "+" is not optional.

The IRAC+ Structure (Canonical)

IRAC+ = Context → Issue → Rule → Application → Conclusion

This order is deliberate.

1. **Context (The +)**

Context answers:

- **Person**: Who provided the information?
- **Place**: Where in the organization or system did this occur?
- **Situation**: What was happening when the issue surfaced?
- **Constraints**: Policy, legal, technical, operational limits
- **Artifacts**: Screens, logs, reports, forms, systems

Context anchors reality.

Without it, everything that follows is fragile.

2. **Issue**

The Issue states:

- What is broken, unclear, risky, or inefficient
- Why it was identified as an issue
- What decision hinges on resolving it

Issues should be:

- Specific
- Observable
- Neutral in tone

Not:

- Accusatory
- Solution-biased
- Hypothetical

3. **Rule**

The Rule answers:

- Which **FDE Rule(s)** apply
- What principle governs this situation
- What constraints must be respected

Rules come from:

- This Rulebook
- Organizational policy
- Legal or regulatory boundaries

The Rule is **not** a conclusion.

It is the lens through which the issue must be viewed.

4. **Application**

Application is where engineering judgment lives.

It explains:

- How the rule interacts with reality
- Where reality aligns with the rule
- Where it conflicts
- What tradeoffs exist

This section must:

- Reference observed facts
- Acknowledge uncertainty
- Avoid premature certainty

Application is the difference between theory and deployment.

5. Conclusion

The Conclusion proposes:

- A fix
- A next step
- An experiment
- Or a decision to defer

Conclusions should be:

- Actionable
- Bounded
- Revisitable

They are **not final truths**.

They are the best move given current information.

Before IRAC+: Raw Field Notes

Collected over three days from meetings, Slack, email, and a shared doc.

Monday — Operations Sync

- "Sensor data looks fine in staging."
- "Field teams say alerts don't match what they're seeing."
- "Probably a calibration issue?"
- "We shipped the new workflow last Friday."
- "Ops says they followed the runbook."

Slack Thread — #deployment

9:14 AM
"Anyone else seeing duplicate alerts from Zone C?"
9:16 AM
"That's expected behavior, I think."
9:22 AM
"No, that shouldn't happen unless the fallback logic fires."
9:31 AM
"Fallback only triggers if sensor confidence < 0.7."
9:45 AM
"Confidence is below 0.7 in the field. Always."

Email — From Vendor

"Our hardware is operating within spec.
Any downstream interpretation issues are likely software-side."

Meeting Notes — Engineering Review

- Confidence threshold set during pilot
- Pilot environment ≠ field environment
- No one owns threshold changes
- Ops assumed alerts were informational
- Engineering assumed alerts were actionable
- Escalation path unclear
- No one sure who approved final workflow

Open Questions (Unanswered)

- Is the system malfunctioning, or behaving as designed?
- Are alerts warnings, or commands?
- Who is allowed to change thresholds?
- Is this a sensor problem, a model problem, or a workflow problem?
- Should the system be rolled back?

Result:

Lots of activity.
No decision.
No ownership.
No shared understanding.

After IRAC+: Structured Decision Record

IRAC+ FIELD MEMO — Deployment #147

Issue

Are duplicate alerts from Zone C evidence of a system malfunction requiring rollback, or the result of a workflow mismatch between engineering assumptions and field conditions?

Rule

1. Alerts must map to a single, clearly defined operational action.
2. Thresholds that change alert meaning require explicit ownership.
3. A system is considered "failing" if it creates consistent operator confusion, even if it is technically correct.

Analysis

- Sensor hardware is operating within specification (vendor confirmation).
- Model confidence in the field is consistently below 0.7 due to environmental noise not present in pilot conditions.
- This causes fallback logic to trigger continuously.
- Engineering assumed fallback alerts were actionable signals.
- Operations interpreted fallback alerts as informational warnings.
- No role was assigned ownership of threshold calibration post-pilot.

As a result:

- The system behaved exactly as designed.
- The workflow behaved incorrectly in the field.

This is not a sensor failure or a model failure.

It is a **workflow authority failure.**

Conclusion

Do not roll back the system.

Instead:

- Redefine fallback alerts as informational only.
- Assign threshold ownership to Field Operations, not Engineering.
- Update the runbook to explicitly state when action is required.
- Schedule a threshold recalibration using live field data.

System remains deployed with modified interpretation rules.

+ (Field Reality Addendum)

Known Constraints

- Environmental conditions vary widely by zone.
- Field operators cannot interpret probabilistic confidence scores during live operations.

Unknowns

- Whether long-term confidence drift will worsen.
- Whether additional fallback states are required.

Next Evidence to Collect

- 7 days of post-change alert logs.
- Operator response timing before vs. after clarification.
- Number of escalations triggered per zone.

Owner

- Field Operations Lead (not Engineering)

Outcome Preserved:

Not just what decision was made — but why, under what constraints, and who owns the next move.

Why IRAC+ Scales

IRAC+ works because it is:

- Lightweight
- Auditable
- Transferable
- Iterative

It supports:

- Single issues
- Multi-issue investigations
- Long-running deployments
- Engineer handoffs
- Post-mortems without blame

Most importantly, it preserves **reasoning**, not just outcomes.

IRAC+ as a Field Discipline

Forward Deployed Engineers use IRAC+:

- After interviews
- After failures
- After surprises
- After "that's weird" moments

Not everything deserves code.

Everything deserves documentation.

IRAC+ ensures that:

- Hard-won insight is not lost
- The organization learns cumulatively
- Fixes improve over time
- Engineers can safely disagree later

The Standard Going Forward

From this point onward, when this book refers to:

- "Document the issue"
- "Capture the insight"
- "Leave a record"
- "Enable continuity"

It means **IRAC+.**

Not notes.
Not tickets.
Not memory.

Structured reasoning, anchored in context.

Transition Forward

IRAC+ defines **how Forward Deployed Engineers think and record reality**.

The next question is just as important:

How do they extract accurate information in the first place—especially when they have no authority?

That requires a separate discipline.

ELICITATION WITHOUT AUTHORITY
How Forward Deployed Engineers Extract Facts in the Field

The Core Problem: Information Without Power

Forward Deployed Engineers operate in environments where:

- They do not manage the people they interview
- They cannot compel compliance
- They lack formal authority
- They inherit mistrust from past failures

And yet, they are expected to understand:

- What actually happens
- Why it happens
- Where systems break
- Which fixes will survive reality

This creates a fundamental constraint:

The FDE must extract accurate information without authority, hierarchy, or pressure.

That is not a soft skill.

It is a discipline.

Why Naïve Questioning Fails

Most engineers are trained to ask questions in one of two flawed ways:

1. **Solution-driven questioning**
    - "Would it help if we automated this?"
    - "Is the problem that the system is slow?"
    - "What feature do you want?"
2. **Confirmation-seeking questioning**
    - "This usually works, right?"
    - "That's not a big issue, is it?"
    - "Everyone follows this process?"

Both approaches contaminate the signal.

They cause stakeholders to:

- Mirror language
- Minimize problems
- Skip edge cases
- Optimize for politeness

The result is clean requirements and broken systems.

Elicitation Is Not Interviewing as Commonly Practiced

Elicitation is not:

- Casual conversation
- Requirements gathering
- Brainstorming
- Status updates

Elicitation is the disciplined extraction of **accurate, decision-relevant information** from humans operating inside imperfect systems.

It assumes:

- Memory is fallible
- Incentives are misaligned
- Confidence is unreliable
- Truth emerges gradually

Forward Deployed Engineers must therefore **engineer the conversation itself.**

The Non-Adversarial Mandate

Effective elicitation follows a non-negotiable principle:
Accuracy collapses under pressure.

When people feel:

- Judged
- Rushed
- Exposed
- Trapped

They stop telling the truth—even unintentionally.

Forward Deployed Engineers must create conditions where:

- Speaking freely is safe
- Uncertainty is acceptable
- Contradictions are explored, not punished
- "I don't know" is allowed

This is not about kindness.
It is about signal integrity.

What the FDE Is Actually Doing

During elicitation, the Forward Deployed Engineer is quietly:

- Testing mental models
- Surfacing hidden rules
- Detecting inconsistencies
- Identifying tacit knowledge
- Mapping informal workflows
- Locating decision boundaries

This work cannot be rushed.

Silence is often more informative than questions.
Details matter more than summaries.
Exceptions matter more than the happy path.

Authority Poisons Signal

Formal authority distorts information in predictable ways:

- Subordinates sanitize answers
- Managers defend decisions
- Teams optimize for appearance
- Problems disappear on paper

Forward Deployed Engineers avoid this by:

- Positioning themselves as learners
- Deferring judgment
- Asking for stories, not opinions
- Requesting walkthroughs, not explanations

This posture is deliberate.

It keeps the signal clean.

Why This Is a Technical Skill

Elicitation is often mislabeled as "communication."

In reality, it is closer to:

- Debugging
- Root-cause analysis
- Systems tracing
- Failure investigation

It requires:

- Hypothesis control
- Bias awareness
- Structured sequencing
- Documentation discipline

Which is why it pairs naturally with **IRAC+.**

Elicitation generates raw material.
IRAC+ preserves it.

The Discipline Going Forward

From this point on, when this book discusses:

- Interviews
- Discovery
- Requirements
- Stakeholder conversations

It refers to **elicitation without authority**.

Not persuasion.
Not extraction.
Not manipulation.

Truth-seeking under constraint.

Transition Forward

Understanding *how* to elicit information is necessary—but not sufficient.

Forward Deployed Engineers must also:

- Sequence conversations correctly
- Detect contradictions
- Preserve continuity
- Avoid poisoning memory

That requires a repeatable execution loop.

THE FDE INTERVIEW LOOP

A Repeatable Method for Extracting Accurate Information

Why Ad-Hoc Conversations Don't Scale

Most failed deployments share a quiet root cause:

The engineer talked to the right people, but in the wrong way, in the wrong order, with no durable record.

Forward Deployed Engineers cannot rely on intuition alone.

They need a **repeatable loop** that:

- Preserves signal
- Reduces bias
- Survives handoffs
- Improves over time

The FDE Interview Loop exists to do exactly that.

The FDE Interview Loop (Overview)

The loop has **six stages**, executed sequentially and revisited as needed:

1. Preparation
2. Open Narrative
3. Clarification & Sequencing
4. Contradiction Detection
5. Artifact Validation
6. Documentation & Continuity

Skipping stages degrades accuracy.
Reordering stages contaminates signal.

1. Preparation (Lightweight, Not Rigid)

Preparation is not about designing solutions.

It is about:

- Identifying what you *think* you know
- Writing down assumptions
- Reviewing existing artifacts
- Noting constraints

This step exists to **externalize bias**, not eliminate it.

Forward Deployed Engineers enter conversations knowing they are probably wrong—but also knowing *how* they might be wrong.

2. Open Narrative (Uninterrupted)

The interview begins with an open request:

- "Walk me through how this actually works."
- "Tell me what happens from start to finish."
- "Describe the last time this went wrong."

During this phase:

- The FDE does not interrupt
- The FDE does not correct
- The FDE does not summarize

This is where:

- Mental models surface
- Language patterns appear
- Omissions become visible

Interruptions destroy this signal.

3. Clarification & Sequencing

Only after the narrative completes does the FDE:

- Ask for clarification
- Resolve timeline gaps
- Sequence steps precisely

This phase focuses on:

- Order of operations
- Decision points
- Hand-offs
- Waiting states
- Exception paths

The goal is **structure**, not judgment.

4. Contradiction Detection

Contradictions are not mistakes.

They are signals.

Forward Deployed Engineers listen for:

- Inconsistent descriptions
- Conflicting rules
- "We always" followed by "except when"
- Policy vs practice mismatches

Contradictions are explored gently:

- "Earlier you mentioned..."
- "Help me reconcile these two steps..."
- "What happens when those conflict?"

This is where reality becomes visible.

5. Artifact Validation

Words are tested against artifacts:

- Screens
- Logs
- Forms
- Reports
- Tickets
- Spreadsheets

Artifacts answer questions people cannot.

When artifacts disagree with narratives, the artifact wins—until proven otherwise.

This step prevents:

- Memory drift
- Social desirability bias
- Retrospective rationalization

6. Documentation & Continuity

Immediately after the interview, the FDE documents findings using **IRAC+**:

- Context captured
- Issues stated
- Rules applied
- Reasoning preserved
- Conclusions bounded

This documentation enables:

- Handoffs
- Re-evaluation
- Audits
- Learning over time

Without this step, insight decays.

Why This Loop Works

The FDE Interview Loop:

- Separates signal from noise
- Prevents premature solutions
- Surfaces tacit knowledge
- Respects human limits
- Scales across domains

It transforms conversations from anecdotes into **engineering inputs**.

Common Failure Modes (and Why They Happen)

- **Jumping to solutions** → discomfort with ambiguity
- **Over-structuring early** → anxiety about time
- **Ignoring contradictions** → social pressure
- **Skipping artifacts** → false confidence
- **Delayed documentation** → cognitive decay

The loop exists to counter these instincts.

The Discipline Forward Deployed Engineers Adopt

Forward Deployed Engineers treat interviews as:

- A system
- A process
- A repeatable experiment

Not as casual conversations.

This discipline is what allows them to operate credibly without authority—and still extract issues and facts.

Transition Forward

Even with a strong interview loop, one risk remains:

The engineer accidentally leads the witness, shaping reality instead of observing it.

That failure mode deserves its own rule.

COMPARATIVE INTERVIEWING FRAMEWORK

How the FDE Method Is Derived—and What It Explicitly Rejects

Why This Comparison Matters

Forward Deployed Engineers borrow rigor from multiple disciplines—but they are not practitioners of those disciplines.

Interviewing has a long history in:

- Military intelligence
- Civil adjudication
- Investigative policing
- Academic research

Each tradition solved a different problem under different constraints.
This section exists to do two things:

1. **Establish legitimacy** — the FDE method is not improvised
2. **Set boundaries** — what FDEs take, and what they explicitly reject

The Three Traditions That Inform FDE Practice

The FDE interviewing method is **derived**, not invented. It draws selectively from three established traditions.

1. Military Intelligence (Preparation & Discipline)

What it contributes

- Careful preparation
- Objective-driven questioning
- Sequencing over improvisation
- Patience under ambiguity

What it solves

- Extracting accurate information when stakes are high
- Avoiding distraction by noise or emotion
- Maintaining discipline under pressure

What FDEs explicitly reject

- Detention contexts
- Coercive authority
- Power asymmetry
- Compliance-driven questioning

The FDE borrows structure, not posture.

2. Civil Adjudication (Non-Adversarial Elicitation)

What it contributes

- Neutral tone
- Open-ended narratives
- Clarification without accusation
- Accuracy over speed
- Documentation discipline

What it solves

- Eliciting testimony without intimidation
- Reducing bias introduced by the interviewer
- Preserving credibility over time

What FDEs explicitly reject

- Legal judgment
- Determinations of guilt or fault
- Procedural formalism
- Outcome enforcement

The FDE borrows neutrality, not authority.

3. Investigative Interviewing (Rapport & Contradiction Resolution)

What it contributes

- Rapport-building
- Narrative-first sequencing
- Attention to inconsistencies
- Separation of fact from interpretation

What it solves

- Getting people to talk freely
- Surfacing contradictions without confrontation
- Understanding how stories change over time

What FDEs explicitly reject

- Accusatory framing
- Deceptive tactics
- False evidence
- Confession-seeking
- Psychological manipulation

The FDE borrows **rapport**, not manipulation.

What emerges from these conversations is later stabilized using IRAC+, so field insight becomes an auditable decision rather than an opinion.

■ CASE FILE

When the System "Worked" — Until It Didn't

Context

A large organization approved a new internal system after months of demos, reviews, and stakeholder sign-off.
Leadership believed the rollout was low risk.
The dashboards were green.
The vendor was confident.
The operators were silent.

The Trap

The system was evaluated through:

- Slide decks
- Success metrics chosen by leadership
- Pilot users who volunteered
- Scenarios that avoided edge cases

No one asked how work actually flowed during peak load, exception handling, or failure recovery.

Silence was mistaken for agreement.

The Bad Move

Engineering relied on:

- Confident presenters
- Senior stakeholders
- Clean pilot results

No one sat with frontline operators during real work.
No one followed a task from start to finish.
No one asked who fixed things when they broke.

The system shipped.

Consequence

Within weeks:

- Operators bypassed the system with shadow spreadsheets
- Exceptions piled up outside automation
- Errors surfaced late, downstream, and expensively
- Trust eroded faster than metrics could detect

Leadership asked why adoption was "lagging."
Engineering asked why requirements were "misunderstood."

Both were asking the wrong question.

The Fix

A Forward Deployed Engineer embedded directly into daily operations.

They:

- Shadowed operators during peak hours
- Watched workarounds form in real time
- Documented exceptions instead of suppressing them
- Re-ran the evaluation using real workflows, not demos

The system was re-scoped.
Automation was narrowed.
Human override paths were formalized.
Ownership was reassigned.

Adoption followed.

The Rule

A system that looks correct in review but fails in practice was never correct.

If you don't observe real work under real conditions,
you are validating narratives — not systems.

The FDE Position (Explicitly Defined)

Forward Deployed Engineers operate under a distinct mandate:

Extract accurate, decision-relevant information without coercion, deception, or authority.

This mandate is narrower than law enforcement and broader than requirements gathering.

It exists specifically for:

- Ambiguous systems
- Incomplete documentation
- High-stakes operations
- Long-lived deployments

A Clear Comparison

Dimension	Military	Civil	Investigative	FDE
Authority	High	Institutional	Situational	**None**
Goal	Intelligence	Testimony	Resolution	**Accuracy for deployment**
Posture	Command	Neutral	Rapport	**Learner**
Methods	Structured	Non-adversarial	Narrative	**Selective synthesis**
Coercion	Possible	Prohibited	Historically mixed	**Explicitly prohibited**
Outcome	Action	Decision	Confession	**Fix-forward execution**

This table exists to prevent misuse.

Why Explicit Rejection Is Necessary

Without clear boundaries, methods drift.

Engineers under pressure may:

- Lead questions unintentionally
- Frame answers prematurely
- Signal preferred outcomes
- Optimize for speed over accuracy

By stating what is **out of bounds**, the FDE protects:

- The stakeholder
- The organization
- The engineer
- The deployment itself

Ethical Discipline as a Technical Requirement

Ethics here are not philosophical.

They are **functional**.

Coercion corrupts signal.
Deception distorts memory.
Authority suppresses truth.

Forward Deployed Engineers reject these not out of principle alone—but because they break systems.

The Resulting Method

The FDE method is therefore:

- Structured but non-coercive
- Patient but decisive
- Neutral but accountable
- Rigorous without being adversarial

It exists to **serve reality**, not dominate it.

Transition Forward

With elicitation, interviewing, and documentation defined, we now turn to a crucial operational rule:

What happens when engineers accidentally influence the very reality they are trying to observe?

That failure mode must be addressed directly.

Chapter 13 — Failure in the Field

"Failure is not the end.
It is the first teacher—
and its lessons are written in scars."
— Forward Deployed Engineer Rule Book

Failure is not optional. In the field, things break: systems, processes, plans, sometimes even trust. The difference between an FDE and everyone else is what you do after it breaks.

Failure is raw data. It shows you what reality thinks of your design. A failed deployment is not the end of the story; it is the beginning of understanding.

Do not hide failure. Study it. Share it. Fix it in daylight. The field respects honesty more than spin.

An FDE who refuses to fail refuses to learn.

Common FDE Failure Modes (and Early Warning Signs)

A Diagnostic Checklist for Field Work Under Pressure

Why This Checklist Exists

Forward Deployed Engineers rarely fail all at once.

They fail **gradually**, through small, understandable deviations that feel reasonable in the moment.

This checklist exists to help FDEs pause and ask:

"*Which failure mode am I drifting toward right now?*"

It is not a judgment tool.
It is a **self-diagnostic**.

How to Use This Page

- Review it **when something feels off**
- Revisit it **before scaling**
- Use it **after setbacks**, not just successes
- Treat multiple warning signs as compounding risk

If you recognize yourself in more than one category, **slow down**.

Discovery Failure Modes

What Goes Wrong:
The FDE unknowingly shapes reality instead of observing it.

Early Warning Signs

- You are proposing solutions during interviews
- Stakeholders frequently agree with you quickly
- Conversations feel "too clean"
- Edge cases appear only after deployment
- Artifacts contradict interviews

Typical Root Cause

- Leading the witness
- Overconfidence in early understanding
- Discomfort with ambiguity

Corrective Action

- Reset interviews
- Return to open narratives
- Re-validate against artifacts
- Re-document using IRAC+

Documentation Failure Modes

What Goes Wrong:

Insight exists only in the engineer's head.

Early Warning Signs

- Notes are informal or scattered
- Decisions cannot be easily revisited
- Handoffs feel risky
- "We'll remember why later"
- Context is missing from tickets or docs

Typical Root Cause

- Time pressure
- Over-reliance on memory
- Treating documentation as overhead

Corrective Action

- Enforce IRAC+
- Capture context explicitly
- Document same day, not later

Execution Failure Modes

What Goes Wrong:
A promising PoC is mistaken for a production-ready system.

Early Warning Signs

- Success is measured by demos
- Real data is deferred
- Error handling is "Phase 2"
- Monitoring is an afterthought
- Users adapt around the system

Typical Root Cause

- Pressure to show progress
- Misaligned incentives
- Fear of slowing down

Corrective Action

- Re-scope the PoC
- Expose failure modes intentionally
- Delay scaling until reality pushes back

Organizational Failure Modes

What Goes Wrong:
The FDE becomes a buffer for dysfunction instead of a catalyst for clarity.

Early Warning Signs

- You translate between teams constantly
- Decisions are deferred upward indefinitely
- No one owns outcomes post-handoff
- Political considerations override reality
- You are asked to "just make it work"

Typical Root Cause

- Missing executive sponsorship
- Unclear ownership
- Cultural resistance to surfacing failure

Corrective Action

- Escalate constraints explicitly
- Document blockers
- Refuse silent heroics

Ethical & Boundary Failure Modes

What Goes Wrong:
Access expands faster than justification.

Early Warning Signs

- "Temporary" access becomes permanent
- Data is reused across contexts casually
- Curiosity justifies access
- Documentation lags behind access
- You feel uneasy explaining your access later

Typical Root Cause

- Informal trust without guardrails
- Lack of explicit boundaries
- Pressure to move fast

Corrective Action

- Re-scope access
- Document justification
- Re-establish minimum-necessary principle

Personal Failure Modes (Burnout & Role Collapse)

What Goes Wrong:
The FDE becomes indispensable—and fragile.

Early Warning Signs

- You are the only one who understands the system
- Progress depends on your presence
- You bypass process "to save time"
- You feel responsible for everything
- You cannot step away safely

Typical Root Cause

- Over-identification with the role
- Lack of handoff discipline
- Rewarding heroics over sustainability

Corrective Action

- Force documentation
- Train a replacement
- Design yourself out of the loop

The Meta-Failure: Ignoring the Signs

The most dangerous failure mode is recognizing these signals—and proceeding anyway.

Forward Deployed Engineers are hired not to be fast, clever, or heroic.

They are hired to be **correct under pressure**.

This checklist exists to help you stay that way.

Transition Back to the Book

Failure is not the opposite of progress.

Unexamined failure is.

The next section explains how Forward Deployed Engineers **fix forward**—without denial, blame, or reset fantasies.

Chapter 14 — Debugging Under Fire

"When systems collapse,
so do illusions.
What remains is truth—and your response."
— Forward Deployed Engineer Rule Book

Debugging in a lab is methodical. Debugging in the field is survival. Systems crash, alarms blare, and people look to you for answers in real time.

Palantir FDEs learned this in Army command tents, patching code mid-exercise while generals waited. This is debugging under fire: grit, presence, judgment — not Jira tickets.

In these moments, excuses vanish. There is no room for blame or theory. You work with the tools you have, the people beside you, and the time left on the clock.

Fire reveals more than failure ever could. It shows you how your system behaves under pressure, how your teammates respond, and how you hold up when the margin for error is gone.

The FDE does not fear the fire. They respect it — and learn from every burn.

Chapter 15 — The Human Factor

"Machines falter. Networks fail.
The human element
is the last redundancy."
— Forward Deployed Engineer Rule Book

Every system has a human layer. Ignore it, and you miss half the system. Technology is brittle. People are adaptable. Partner with them by capturing workarounds as design input: every sticky note, logbook, and shortcut is a requirement for the next iteration.

Technology is brittle. People are adaptable. That adaptability is both strength and risk — it keeps things running, but it also hides flaws that must eventually be fixed.

The FDE respects the human factor. Partner with it, don't fight it. Build systems that support people, not systems that demand people compensate for their weaknesses.

Chapter 16 — Fix Forward

"You cannot rewind the clock.
You can only rebuild tomorrow."
— Forward Deployed Engineer Rule Book

When things break — and they will — you face a choice: dwell on the past or build for the future. The FDE always fixes forward.

Fix forward means learning from failure without being paralyzed by it. It means applying the lesson to the next version, the next deployment, the next opportunity.

The field does not wait for perfection. It rewards momentum. Move forward with clarity, humility, and speed.

To fix forward is not to forget — it is to honor the past by refusing to repeat it.

🗀 CASE FILE

The System That Almost Shipped Wrong

Context

A team was preparing to deploy an automated decision system into a live environment.
The rollout date was set.
The metrics looked strong.
The integration tests passed.

From a distance, the system appeared ready.

The Near-Miss

During a final field walk-through, a Forward Deployed Engineer noticed a pattern that did not appear in logs or dashboards.

Operators hesitated before confirming certain outputs.
They double-checked results manually.
Some delayed execution until a second person verified the outcome.

Nothing was broken.
Nothing failed loudly.

But something felt wrong.

The Bad Assumption

The system had been evaluated on:

- Historical data
- Clean test cases
- Average workloads

It assumed:

- Inputs arrived in a stable order
- Edge cases were rare
- Human review was optional

None of those assumptions held in the field.

The Intervention

Instead of rolling back or halting the launch, the FDE applied a fix-forward approach.

They:

- Replayed real workflows end-to-end with operators present
- Captured hesitation as signal, not resistance
- Identified decision points where automation outpaced trust
- Narrowed the automated scope instead of expanding it
- Added explicit pause and override paths

The deployment was delayed — not canceled.

The Outcome

The system shipped later, smaller, and safer.

Because the near-miss was treated as information:

- No incident occurred
- No public rollback was required
- Trust increased instead of eroded
- The organization learned without paying full failure cost

The Rule

Fixing forward does not mean shipping faster.
It means shipping wiser.

Near-misses are warnings.
Ignore them, and the field will teach the lesson louder.

DATA ACCESS, OWNERSHIP, AND BOUNDARIES

What Forward Deployed Engineers May Touch—and What They Must Not

Why Boundaries Are a Technical Requirement

Forward Deployed Engineers operate close to sensitive systems:

- Personal data
- Operational records
- Logs and audit trails
- Internal communications
- Decision artifacts

Proximity creates power.

Power without boundaries creates risk.

This section exists to make those boundaries explicit—not as legal advice, but as **engineering discipline.**

Access Is Not Entitlement

Being embedded does not mean having unlimited access.

Forward Deployed Engineers treat access as:

- Purpose-bound
- Time-bound
- Scope-bound

Every dataset, system, or artifact must answer a simple question:

What decision does this access enable right now?

If the answer is unclear, access is premature.

Principle: Minimum Necessary Reality

Forward Deployed Engineers follow a minimum-necessary principle:

- Access only what is required to understand the issue
- Avoid broad pulls "just in case"
- Prefer sampling over extraction
- Escalate access deliberately

This protects:

- Stakeholders
- The organization
- The engineer
- The deployment itself

It also builds trust.

Public Data, Private Context

In public-sector environments, some data may be legally accessible through public mechanisms.

That does not make it ethically interchangeable with internal operational data.

Forward Deployed Engineers distinguish between:

- **Legal access** — what is permitted
- **Operational access** — what is appropriate
- **Contextual access** — what is necessary for this decision

Legal permission does not eliminate ethical responsibility.

Ownership of Work vs Ownership of Insight

A frequent source of confusion is ownership.

Forward Deployed Engineers recognize three distinct categories:

1. **Code and artifacts**
   - Typically owned by the organization
   - Must follow internal policy
   - Should be documented and transferable
2. **Process insight**
   - Emerges from field observation
   - Should be preserved for continuity
   - Must be documented responsibly
3. **Judgment**
   - Resides with the engineer
   - Must be defensible
   - Must be explainable

Confusing these categories leads to conflict.

Documentation as a Boundary Tool

IRAC+ is not just a reasoning framework.

It is a **boundary mechanism**.

By documenting:

- Context
- Scope
- Rules applied
- Assumptions made
- Conclusions reached

The Forward Deployed Engineer:

- Limits misinterpretation
- Prevents scope creep
- Enables review
- Protects future operators

Undocumented access is indistinguishable from abuse.

What Forward Deployed Engineers Must Avoid

Regardless of environment, Forward Deployed Engineers must not:

- Circumvent controls "temporarily"
- Access data unrelated to the active issue
- Retain copies without authorization
- Reuse data across contexts casually
- Treat curiosity as justification

These behaviors undermine trust—even if intentions are good.

Ethics as a Deployment Constraint

Ethical boundaries are not philosophical abstractions.

They are constraints that shape system design.

Ignoring them leads to:

- Regulatory exposure
- Reputational harm
- Organizational backlash
- Project termination

Forward Deployed Engineers integrate ethics early—when change is still cheap.

The Professional Standard

The Forward Deployed Engineer's credibility rests on a simple expectation:

If someone audits this work later, the reasoning will still make sense.

Boundaries make that possible.

Transition Forward

With boundaries defined, the next challenge is scale.

How do Forward Deployed Engineers expand impact without losing:

- Context
- Ownership
- Ethics
- Trust

That question defines the final section.

Part IV – The Future of Engineering Work

Chapter 17 – AI as a Teammate

"AI is not prophecy.
It is power in your hands,
waiting to be directed with purpose."
— Forward Deployed Engineer Rule Book

AI is powerful, but not prophetic. Left unchecked, it amplifies noise. Guided by an FDE — with dual credentials in code and domain — AI becomes a teammate: strong, tireless, but needing clear direction.

Left unchecked, AI amplifies noise. Directed well, it amplifies signal. The difference is not in the model, but in the human who sets the aim.

The field will never be automated away. It will be reshaped. And the FDE will stand at the intersection, guiding AI to serve people rather
than replace them.

MODERN SOFTWARE & SELF-IMPROVING SYSTEMS

Why Forward Deployed Engineers Multiply Impact

Software Is the Lever—Judgment Is the Fulcrum

Forward Deployed Engineers are not valuable because they write code.

They are valuable because they decide **what code should exist at all.**

Modern software—especially AI-enabled systems—can amplify effort dramatically. But amplification cuts both ways. Poorly chosen automation multiplies waste just as efficiently as well-chosen automation multiplies value.

The difference is judgment.

Why "Self-Improving" Systems Fail Without Field Ownership

Many systems are described as self-improving because they:

- Retrain on new data
- Incorporate feedback loops
- Adjust thresholds
- Optimize metrics

In practice, most of these systems:

- Drift silently
- Optimize the wrong objective
- Encode workarounds
- Mask failure behind dashboards

Without field ownership, feedback loops become **echo chambers**. Forward Deployed Engineers close the loop by:

- Validating feedback sources
- Inspecting failure cases
- Updating objectives when reality shifts
- Preventing metric gaming

Self-improvement is not automatic.
It is supervised.

Why Python, Scripts, and "Small Tools" Matter More Than Platforms

In the field, leverage often comes from:

- Small scripts
- One-off tools
- Glue code
- Data checks
- Automation of boring steps

These tools:

- Reduce cognitive load
- Eliminate human error
- Shorten feedback cycles
- Free time for judgment

Forward Deployed Engineers prefer **simple, inspectable systems** over large platforms early on.

Complexity is earned—not assumed.

Automation as Hypothesis, Not Commitment

FDEs treat automation as a hypothesis:

- "If we automate this step, does value increase?"
- "If we remove this manual check, what breaks?"
- "If this model fails, who notices?"

This mindset prevents:

- Premature scaling
- Fragile dependencies
- Irreversible architecture

Automation is reversible when designed by someone embedded in the work.

Feedback Loops That Actually Work

Effective self-improving systems share three properties:

1. **Ground-truth feedback** (not proxy metrics)
2. **Human override paths**
3. **Clear ownership**

Forward Deployed Engineers ensure that:

- Feedback is representative
- Failures surface quickly
- Humans remain in the loop where judgment is required

The goal is not autonomy.
The goal is **reliable assistance**.

Why FDEs Prefer Incremental Intelligence

Large model upgrades promise step-function improvements.

Field reality prefers:

- Incremental gains
- Narrow wins
- Local optimization
- Measurable impact

Forward Deployed Engineers bias toward:

- Systems that can be understood
- Failures that can be diagnosed
- Improvements that can be validated

This conservatism is a strength.

The Engineering Shift This Enables

When software becomes:

- Easier to write
- Faster to deploy
- Cheaper to iterate

The bottleneck shifts decisively to:

- Problem selection
- Constraint identification
- Trust calibration
- Ethical boundary setting

These are not abstract concerns.

They are field decisions made daily by Forward Deployed Engineers.

Transition Forward

Modern tools give Forward Deployed Engineers extraordinary leverage.

The final question is how to scale that leverage **without losing alignment**, ownership, or ethics.

That is the challenge of scale—and it demands restraint.

Chapter 18 – Scaling the Field

"Scale without empathy
is collapse on a larger stage.
Grow wisely, or fail loudly."
— Forward Deployed Engineer Rule Book

Scaling a solution is easy in theory, brutal in practice. What works for one team may fail spectacularly for a thousand.

The FDE knows scale is not just about throughput or servers. It is about people, trust, and adoption. A fix that ignores human reality will collapse the moment it grows.

Scale wisely. Test in the wild. Never assume that multiplying a solution will multiply its success. Sometimes, it multiplies its flaws.

Chapter 19 – Global Engineering

"Borders divide maps,
but problems cross them freely.
The field is everywhere."
— Forward Deployed Engineer Rule Book

Engineering does not stop at borders. Supply chains, data flows, and human migration make every system global. The Immigration Case Study demonstrates this: agencies across **DOJ**, **DHS**, **DOS**, **SSA**, and **IRS** reveal fragmentation that only cross-border immersion can resolve.

The FDE must think beyond the local. What works in one region may break in another. What is legal in one country may be forbidden in the next.

Global engineering means building with cultural awareness, regulatory foresight, and humility. The problems of tomorrow will not fit neatly inside national boundaries. Neither can the solutions.

Chapter 20 – Rule Makers and Rule Breakers

"Rules steady the hand—
until they shackle it.
Know when to follow,
and when to forge new ground."
— Forward Deployed Engineer Rule Book

Rules exist for a reason. They protect, stabilize, and guide. But every rule has an expiration date - a point where it stops enabling and starts constraining.

The FDE must respect rules but not worship them. Know when to follow, when to bend, and when to build anew.

The future belongs to those who can tell the difference between a rule worth keeping and a rule worth breaking.

Chapter 21 – The Future FDE

"The best engineers are apprentices forever.
The field never stops changing,
and neither should you."
— Forward Deployed Engineer Rule Book

The role of the FDE will not stay still. New tools will arrive, new fields will emerge, and the definition of "the problem" will keep shifting.

What will not change is the posture: stepping into the field, listening, learning, and solving side by side.

The future FDE is not the master of all trades. They are the apprentice of every new one. The humility to keep learning is the most durable skill of all.

Part V – Appendices & Case Studies

Appendix Overview – The FDE Toolkit

The appendices are not theory. They are tools.
Each one captures scars, shortcuts, and stories you can carry into the field.

A. Case Studies – Lessons From the Field
Palantir, GitLab, Devon/DOE, A16Z, Immigration.
☛ *The field is the final exam.*

B. Templates – Shortcuts, Not Solutions
Maps that must be broken and rebuilt.
☛ *Templates buy speed. Adaptation buys survival.*

C. Rule Templates – Scars Into Shortcuts
Guidelines forged in failure, tested in fire.
☛ *Every rule template carries blood in its ink.*

D. Career Roadmap – The FDE Path
From student → insider → apprentice → FDE → startup → leader.
☛ *The FDE is not a title. It is a path.*

E. Reading & Training – Sharpening the Edge
Books for perspective. Drills for reflex.
☛ *Apprenticeship never ends.*

APPENDIX A - Case Studies A~F

"Problems do not live in PowerPoint. They live in the field.
Each case study here is drawn from real deployments — where adoption was earned, systems broke, and engineers fixed forward."

Case Study A
Palantir & the Army – Government FDE

Case Study: Palantir & the Army – FDEs in Uniform

Problems do not live in PowerPoint. They live in the field.
When the U.S. Army adopted Palantir, they didn't just buy software — they got forward deployed engineers who lived on base, ate in the chow halls, and patched systems under live fire drills.

Palantir's FDEs weren't consultants. They weren't contractors parachuting in with slide decks. They were embedded, authorized, and accountable. They watched generals slam laptops shut in frustration. They heard operators curse as dashboards froze mid-mission. And instead of filing tickets, they fixed forward — sometimes in the same tent where soldiers were planning the next exercise.

Immersion with Mandate

The difference between Palantir's Army deployments and corporate rollouts was **mandated authority**. Soldiers didn't get to say "we'll adopt later." The order came from the top: you will use this system.

But authority alone wasn't enough. The FDEs had to earn trust on the ground. A system forced on operators without working in their hands collapses. Immersion mattered as much as mandate.

Debugging Under Fire

Exercises are where theory breaks.

One Palantir FDE described patching code while sitting next to a commander in a live war game: "We had seconds, not sprints." The lesson? Field deployments don't wait for sprint cycles. Debugging under fire means grit, presence, and judgment — not Jira tickets.

Tension in the Role

This model created tension.

- **Mandated Authority**: Everyone must use the system.
- **Practical Immersion**: Everyone will abandon the system if it fails in the moment.

The FDE stood in the middle. They were translators between command intent and operator reality. They embodied the paradox of the role: rules imposed from above, systems fixed from below.

Lessons for FDEs

1. **Immersion reveals truth** — Orders don't change friction; presence does.
2. **Fix forward, even under mandate** — Rolling back wasn't an option. Lives, trust, and credibility were on the line.
3. **Authority without empathy collapses** — You can force software into the field, but only immersion keeps it alive there.

Why It Matters

Palantir's Army deployments crystallized what "Forward Deployed Engineer" really means:

- Not an engineer in the office.
- Not a consultant delivering slides.
- But a translator, builder, and firefighter inside the system where failure hurts.

This is not just history — it is a model. Government FDEs still wrestle with the same tension today: authority vs. immersion. And corporate FDEs must ask themselves: Do we earn adoption, or do we assume it?

Case Study A – Palantir & the Army: FDEs in Uniform

Problems do not live in reports. They live in command tents.

When the U.S. Army adopted Palantir, they didn't just buy software — they got engineers who lived on base, debugged under fire, and stood between generals and failing dashboards. Orders mandated adoption, but only immersion earned belief.

Emerging Solutions

- Real-time battlefield data feeds from drones and satellites.
- AI copilots embedded into command tents for live decision support.
- Secure mesh networks resilient to jamming and outages.
- Soldier-wearable sensors streaming health and location data.

Lessons for FDEs

1. Immersion beats dashboards — only field presence keeps systems alive under stress.
2. Secure comms are lifelines — mesh networks and AI copilots collapse without trust.
3. Sensors expand the battlefield view — but require seamless integration, not extra burden.
4. Fix forward under fire — iteration in real time is adoption's only guarantee.

Case Study B
GitLab Adoption – Corporate FDE

Case Study: GitLab Adoption – FDEs as Accelerators

Technology adoption is never just technical.
When enterprises bought GitLab, they weren't only buying version control or CI/CD pipelines. They were buying forward deployed engineers who could make adoption stick. Without them, GitLab risked becoming just another unused license. FDEs inside GitLab understood this: **adoption was the real battlefield.** If the client didn't integrate GitLab into daily work, the deal was dead. The code could be perfect — the people still had to use it.

Adoption as a Human Problem
Adoption friction rarely came from the product itself. It came from managers blocking change, legacy workflows that no one wanted to give up, or developers who feared breaking builds. GitLab FDEs weren't there to sell features. They were there to sit beside client teams, run live projects, and prove that the system worked in the mess of real pipelines.

The Role of the Adoption Accelerator
GitLab's FDEs acted as **adoption accelerators**.

- They shadowed teams to watch how code actually flowed.
- They built one-off integrations that smoothed the rough edges.
- They trained engineers in real time, using the client's own projects as the classroom.

Adoption wasn't left to chance — it was engineered, step by step.

Stories as Signals

Each client story became a **signal**.

- A team struggling with GitLab runners = signal to simplify training.
- A manager resisting integration = signal that leadership alignment was missing.
- A client writing workarounds = signal of a product gap that had to be closed.

FDEs treated these stories not as anecdotes but as data. Stories as signals became a principle of deployment.

Lessons for FDEs

1. **Adoption is the real metric** — A license without use is a ghost victory.
2. **Stories reveal signals** — Every frustration is a data point, if you listen.
3. **FDEs turn resistance into trust** — Presence converts skeptics faster than features.

Why It Matters

GitLab's case proves that **corporate FDEs don't only deliver technology — they deliver adoption**.

Without them, even brilliant tools gather dust. With them, enterprises don't just install new software; they change how they work.

The rule is clear: **In corporate deployments, adoption is the mission.**

Case Study B – GitLab Adoption: FDEs as Accelerators

Technology adoption is never just technical.

GitLab's real product wasn't just code hosting. It was adoption — making enterprises actually use the system. FDEs embedded with client teams, turned resistance into trust, and treated stories as signals.

Emerging Solutions

- AI-driven code review copilots for live pair programming.
- Cloud-native CI/CD pipelines scaled across geographies.
- Secure DevOps with compliance checks embedded at commit time.
- Continuous adoption analytics to track trust, not just tickets.

Lessons for FDEs

1. Adoption is the real product — licenses without live use are failures.
2. AI tools amplify adoption — but only if introduced in real workflows.
3. Compliance must live inside the pipeline — not as an afterthought.
4. Stories are signals — every frustration is actionable field data.

Case Study C
Devon Energy & DOE – Hybrid Engineer

■ Case Study: Devon Energy & DOE – Hybrid Engineers in the Field

Some problems need more than software. They need hybrids.

Devon Energy and U.S. Department of Energy projects revealed a truth: pure software engineers couldn't solve field problems alone. Wells, grids, and pipelines had their own physics, their own risks. A forward deployed engineer here had to wear two hats — coder and domain expert.
That's how the **Hybrid Engineer** emerged. A professional who could sit at a rig site or control room, speak the operators' language, and still open a laptop to push code.

From Office to Well Pad
At Devon, hybrid engineers didn't stay in conference rooms. They ran "data bootcamps" on-site, teaching petroleum engineers how to use new analytics systems. They knew the rhythm of field shifts, the smell of crude, the constraints of drilling schedules.
When a pump sensor failed, they didn't wait for tickets — they pulled logs, wrote scripts, and patched workflows before the downtime cost millions.

DOE's Experiment
The Department of Energy went further: embedding hybrid engineers inside national labs and energy sites. Their mission wasn't just code — it was **translation**. Turning scientific results into operational systems, and operational problems into data-ready models.

The lesson was stark: without hybrids, projects died in the gap between science and operations. With hybrids, field teams trusted the system because the engineer spoke their language.

Scaling with Empathy

Hybrid engineers also proved that **scaling with empathy** matters as much as scaling code.
When Devon Energy rolled out new tools across multiple basins, hybrid FDEs built training into the process. They didn't assume every site would adapt the same way. They created playbooks, ran workshops, and respected the culture of each team.
Scaling wasn't copy-paste. Scaling was empathy, applied to operations.

Lessons for FDEs

1. **Hybrid skill sets unlock adoption** — Knowing the field's language earns credibility no code alone can.
2. **Bootcamps beat manuals** — Training in context works; documentation doesn't.
3. **Empathy scales systems** — Solutions that ignore people break faster than servers.

Why It Matters

The Devon/DOE story shows what happens when FDEs stop being pure coders and become hybrids.
In energy, infrastructure, and science-heavy domains, forward deployment requires domain fluency as much as technical skill.

The rule is clear: **When the system has its own physics, the engineer must have two toolkits.**

Case Study C – Devon Energy & DOE: Hybrid Engineers

Some problems need more than software. They need hybrids.

Pure coders couldn't solve field problems in oil and energy. Hybrid engineers — fluent in both petroleum physics and code — earned trust at rigs, ran on-site bootcamps, and built empathy into scaling.

Emerging Solutions

- IoT sensors on wells and rigs feeding real-time digital twins.
- Predictive AI for grid reliability and outage forecasting.
- Robotics for hazardous inspections in remote or dangerous sites.
- Hybrid cloud edge deployments to keep analytics alive offline.

Lessons for FDEs

1. Hybrids unlock trust — dual-credential engineers bridge physics + code.
2. Digital twins give foresight — but adoption requires local training and buy-in.
3. Edge resilience matters — when the grid fails, the model must survive offline.
4. Empathy scales faster than servers — trust multiplies adoption.

Case Study D
A16Z/Carbon Startups – Moat vs. Margin

Case Study: Startups, A16Z, and Carbon – Trading Margin for Moat

Startups don't have time for theory. They trade margin for survival.

Andreessen Horowitz called it out: startups that deploy forward engineers early may burn cash, but they build moats that keep competitors out. The Carbon startup ecosystem — companies fusing AI with physical industries — proved the point.

They hired expensive FDEs, placed them directly in customer sites, and solved problems others only promised to fix. Yes, the burn rate spiked. But in return, customers became locked in, competitors fell behind, and adoption hardened into dependence.

The Moat vs. Margin Trade

Startups live by a brutal math:

- Hire fewer engineers, keep margin high → risk irrelevance.
- Hire FDEs, embed them in the field → build an adoption moat.

Carbon startups chose the latter. They didn't sell features. They sold transformation, delivered by engineers sitting next to operators, redesigning workflows in real time.

Why It Worked

Customers didn't care about pitch decks or roadmaps. They cared about who was in the room when systems failed.

- Competitors showed demos.
- Carbon startups showed FDEs, fixing problems live.

By the time rivals tried to undercut on price, the moat was already built. Customers wouldn't switch — too much trust, too many workflows, too much presence embedded in the field.

Lessons for FDEs

1. **Margin is fuel, not safety** — Spend it early to win trust.
2. **Moats are built by presence** — Competitors can clone features, not relationships.
3. **Startups win by embedding, not promising** — The field is the sales pitch.

Why It Matters

The Carbon and A16Z perspective reframes the role of FDEs: they aren't a cost center; they are a moat.
Every engineer deployed into the field isn't just solving problems —
they are laying bricks in the wall that protects the company's future.

The rule is clear: **In startups, margin is expendable. The moat is priceless.**

Case Study D – A16Z & Carbon Startups: Trading Margin for Moat

Startups don't have time for theory.

Startups burned margin by embedding expensive FDEs directly in customer sites. The cost built moats that competitors couldn't cross. Trust and presence, not features, became the differentiator.

Emerging Solutions

- Climate modeling platforms combining satellite imagery + AI.
- Robotics for carbon capture and industrial monitoring.
- Smart contracts and blockchain registries for carbon credit integrity.
- AI copilots for compliance, safety, and environmental reporting.

Lessons for FDEs

1. Margin is expendable — presence in the field is the true moat.
2. Trust is harder to copy than code — adoption sticks when operators feel ownership.
3. Emerging tools win when embedded — not when pitched in slides.
4. Startups scale by fixing live — not promising future roadmaps.

Case Study E – Immigration: The Ultimate Jigsaw Puzzle

Note: *Unlike other case studies drawn from real deployments, this is a hypothetical systemic redesign exercise. It illustrates fragmentation across embassies, ports, ICE, OPLA, judges, Coast Guard, SSA, and IRS — a puzzle no one agency has solved.*

Case Study E – Immigration: The Ultimate Jigsaw Puzzle

No system is more fragmented (and yet to be solved).

Immigration spans embassies, ports, ICE, OPLA, judges, Coast Guard, SSA, IRS — each disconnected. FDEs see the whole puzzle, not just pieces.

Emerging Solutions
Biometric Smart Glasses: Instant facial capture at land borders and POEs.
Data Sharing: Require Visa Waiver countries to provide passport-linked biometrics and criminal records.
Mobile Verification: Selfie submissions during visa/ESTA applications.
Case Law Harmonization: Resolve "circuit court loopholes" where rulings apply unevenly across states.
Dual Citizenship Gaps: Close exploitation routes in VWPP/ visa processes.

Next-Generation Technologies
Satellites, robots, drones, smart glasses, quantum communication, advanced fiber-optic sensing, and quantum gravity sensors (to name a few) will reshape the enforcement landscape.

- **Quantum Communication**: Built on the no-cloning theorem and quantum entanglement, enabling secure transmission of biometric and database information. This could allow verified identity data to be pushed directly to smart glasses for real-time use.

- **Advanced Fiber-Optic Sensing & Quantum Gravity Sensors**: Together, these could establish a "virtual underground border fence." Tunnels from Mexico to the U.S. — one of the most dangerous smuggling routes — would be detectable via vibrations, density shifts, and gravitational anomalies, exposed long before becoming operational.

Lessons for FDEs

1. Emerging tools demand integration — satellites, drones, and quantum systems only succeed when unified into a field-ready network.
2. Quantum communication makes trust portable — secure data transfer follows the officer into the field.
3. Subterranean threats require new sensing — fiber-optic and quantum gravity tools extend the border underground.
4. Adoption is the mission — even advanced tools fail without operator trust and usability.

Case Study F – Local Governments: Scale in Fragments

Local governments are fragmented — each with its own funding, culture, and systems. Every broken link is an opportunity for redesign. Immigration may be the federal puzzle; counties and cities are the municipal test.

Emerging Solutions

- Cloud-based citizen portals consolidating services (licenses, benefits, permits).
- Mobile-first applications for public-facing workflows.
- Blockchain registries for property, voting, and contracts.
- AI copilots for frontline clerks to cut wait times and reduce error.

Lessons for FDEs

1. Fragmentation is opportunity — every broken process is a redesign chance.
2. Scale with empathy — adoption depends on local trust, not just tools.
3. Mobile-first is non-negotiable — most citizens interact via phones.
4. Small wins multiply — one county's payroll fix becomes a template for a hundred more.

Appendix A – System Design Exercise – Immigration Case Study E

Immigration as the Ultimate Jigsaw Puzzle

Immigration in the United States is the quintessential example of why the Forward Deployed Engineer (FDE) role must exist. Despite decades of reform attempts, no one has ever seen the system end to end. Each branch, each agency, and each location operates like its own puzzle. Together, these puzzles form a vast, shifting mosaic that no single federal employee has ever assembled.

The Problem

- **Fragmentation**: Every agency has its own policies and procedures. A CBP officer trained at LAX cannot immediately transfer to JFK or San Ysidro without new "port training."
- **Incentives**: Federal employees are rewarded for time served, not for solving systemic problems. Longevity guarantees a pension; innovation rarely changes careers.
- **Politics**: Priorities shift with every administration, fueling apathy among already disengaged staff.
- **Partial Vision**: Retired officials cannot provide a big-picture view, because they never rotated beyond their narrow assignments.

The FDE Mandate

To break this stalemate, Congress would appoint a Forward Deployed Engineer for Immigration:
Appointed by Act of Congress, removable only by Congress.

- Reports directly to Congress, not the President or agency heads.

- Neutral third party, like an escrow agent, but for truth-finding.

The FDE Mandate

- To break this stalemate, Congress would appoint a Forward Deployed Engineer for Immigration:
- Appointed by Act of Congress, removable only by Congress.
- Reports directly to Congress, not the President or agency heads.
- Neutral third party, like an escrow agent, but for truth-finding.
- Explicit, documented authority within their assigned scope to remove employees who obstruct cooperation.
- Dual-credentialed: Computer Science/Software Engineering plus J.D. (law degree).

Training Ground
Training as Field Entry: Immigration

The Forward Deployed Engineer's first stop is not a dashboard or a codebase. It is training.
For domestic immigration enforcement and adjudication, this begins at the Federal Law Enforcement Training Center (FLETC). The FDE completes Immigration & Naturalization modules across agencies and departments, including specialized coursework aligned with the roles of Immigration Judges, Asylum Officers, Consular Officers, and DHS personnel responsible for immigration-related IT systems. Training is not treated as onboarding, but as reconnaissance — an opportunity to observe how rules are taught, where interpretations diverge, and where operational reality already strains formal doctrine.

That immersion is then extended through Department of State consular training, primarily delivered via the Foreign Service Institute (FSI). This training prepares Foreign Service Officers for visa adjudication, citizen services, and post management, with intensive focus on U.S. immigration law, fraud prevention, security procedures, and automated systems. Dedicated modules on Immigrant (IV) and Nonimmigrant (NIV) visas, passports, and special services expose how the same statutes are interpreted differently under diplomatic, legal, and operational constraints.

Seen together, FLETC and FSI do not represent redundant education.

They reveal the core challenge of immigration systems: the same laws are taught, interpreted, and enforced through multiple institutional lenses. Even training itself becomes a source of signal — surfacing inconsistencies, latent conflicts, and structural gaps long before software or policy changes are proposed.

Rotations Through the System

The FDE would rotate through every entry, adjudication, and enforcement point, with tours of 30–90 days:

- **Embassies & Consulates** (DOS/FSI): Visa processing.
- **Visa Waiver Program** (CBP/ESTA): Security vetting.
- **National Targeting Center** (CBP/ATS): Risk analysis.
- **Ports of Entry (CBP):** Airports (LAX, JFK), land borders (San Ysidro), seaports (Houston, LA), preclearance sites (Canada, Ireland, **UAE), and territories (Guam, Puerto Rico, CNMI).**
- **Service Centers (USCIS):** Petition adjudication (I-130, I-140).

- **Field Offices (USCIS):** Adjustment of status and naturalization interviews (I-485, N-400).
- **Application Support Centers (USCIS):** Biometrics collection.
- **Asylum Offices (USCIS): Firs**thand asylum claim reviews.
- **Border Patrol (CBP):** Southwest and northern borders, plus checkpoints for "entries without inspection" (EWI).
- **ICE Investigations (HSI):** Criminal enforcement in major cities.
- **ICE Enforcement & Removal (ERO):** Field offices and detention centers.
- **Deportation Officers:** Managing active caseloads in removal proceedings.
- **OPLA:** Litigating before immigration courts.
- **Immigration Judges** (DOJ/EOIR): Presiding over removal proceedings.
- **Coast Guard**: Offshore vessel inspections near U.S. territory.
- **SSA & IRS**: Investigating fraudulent Social Security and taxpayer IDs.

Tools and Authority

The FDE would hold top-level OCSO access to all DHS databases and programs, unrestricted by IT silos. Beyond observation, they would prototype solutions in real time using **Copilot, Gemini CLI, Claude Code,** and **SpeckIt** for spec-driven development. Mock ports of entry at FLETC would allow safe simulation before field deployment.

Emerging Solutions

- **Biometric Smart Glasses**: Instant facial capture at land borders and POEs.
- **Data Sharing**: Require Visa Waiver countries to provide passport-linked biometrics and criminal records.
- **Mobile Verification**: Selfie submissions during visa/ESTA applications.
- **Case Law Harmonization**: Resolve "circuit court loopholes" where rulings apply unevenly across states.
- **Dual Citizenship Gaps**: Close exploitation routes in VWPP/visa processes.

Output

The FDE's findings would be published as a Government Printing Office report (classified and unclassified). This *Immigration Puzzle Report* would provide Congress with the first integrated map of the system—identifying contradictions, gaps, and modernization opportunities.

Why Only an FDE Can Solve This

Consultants and piecemeal reforms lack both authority and access.

Only an embedded, congressionally mandated officer—with field authority, dual credentials, and neutrality—can assemble the complete picture.

Immigration demonstrates the larger truth: without FDEs, no government or corporation can resolve its most complex systemic puzzles.

🗀 CASE FILE

When Authority Replaced Understanding

Context

A public-sector modernization effort launched with full mandate.
Authority was clear.
Access was granted.
Budgets were approved.

The system touched multiple agencies, overlapping jurisdictions, and time-critical decisions affecting real people.

On paper, everything required to succeed was present.

The Failure

Each component worked in isolation.

- Intake systems captured data correctly
- Adjudication tools followed policy
- Enforcement systems executed decisions
- Reporting dashboards showed compliance

But the end-to-end system failed.

Cases stalled between handoffs.
Contradictions accumulated across databases.
Exceptions were resolved locally and forgotten globally.

No single failure triggered an alarm.

The Bad Assumption

The program assumed:

- Authority guarantees alignment
- Policy consistency implies operational consistency
- If each subsystem works, the whole system works

None of these assumptions survived contact with reality.

Operators adapted silently.
Workarounds filled the gaps.
The system appeared compliant while becoming incoherent.

The Cost

The cost was not a single incident.

It was:

- Compounding delays
- Conflicting decisions across offices
- Appeals driven by inconsistency, not error
- Loss of trust from both operators and the public

Leadership received reports.
Operators absorbed the friction.

The system drifted.

The Missed Opportunity

Access was used to observe components — not workflows.
Authority enforced compliance — not coherence.
Modernization focused on tools — not handoffs.

No one owned the system between systems.

The Lesson

Public-sector systems do not fail because of lack of authority.
They fail because authority is mistaken for understanding.

The Rule

Mandate opens doors.
Only immersion reveals reality.

Without an end-to-end field view,
power accelerates fragmentation instead of fixing it.

📁 CASE FILE

Recovery Without Rewriting History

Context

A public-sector system failed visibly.

The failure was external, measurable, and undeniable.
Cases were delayed.
Decisions conflicted.
Public trust was damaged.

Leadership pressure followed immediately:

- Calls for speed
- Calls for accountability
- Calls for a new system

The instinct was to replace — not to understand.

The Risk

After public failure, organizations default to:

- Narrative repair instead of system repair
- New tools instead of new understanding
- Structural resets that erase evidence

The danger was not repeating the failure.
The danger was **losing the only clear signal the system had ever produced**.

The FDE Move

A Forward Deployed Engineer was embedded after the failure — not to defend it, and not to excuse it.

Their mandate was narrow:

- Do not justify the past
- Do not promise transformation
- Do not redesign prematurely

Instead, they focused on reconstruction.

The Work

The FDE:

- Replayed failure timelines end-to-end
- Reconstructed decisions using IRAC+ artifacts
- Identified where judgment was forced without context
- Mapped where operators compensated for system gaps
- Distinguished policy failure from implementation failure

Nothing was hidden.
Nothing was softened.

The failure remained visible — but explainable.

The Recovery

Recovery did not begin with a new system.

It began with:

- Clear ownership of decisions
- Explicit boundaries on automation
- Narrow fixes applied where failure occurred
- Shared understanding across agencies

Only after coherence was restored did modernization resume.

Trust returned slowly — but it returned.

The Outcome

The organization did not erase the failure.

It learned from it.

Future decisions referenced the failure openly.
New deployments were evaluated against it.
The system improved without pretending the past never happened.

The Rule

Recovery is not reversal.
It is reconstruction.

You do not regain trust by hiding failure.
You regain it by making failure legible — and impossible to repeat.

What Good Looks Like After 12 Months

How to Recognize a Successful Forward Deployed Engineering Engagement

Why This Section Exists

Forward Deployed Engineers are often evaluated incorrectly.

They are praised for:

- Speed
- Visibility
- Heroics
- Constant presence

Those signals are misleading.

This section defines success not by how essential the FDE feels—but by how **durable the system becomes without them**.

The 12-Month Question

After a year, a successful FDE engagement can answer “yes” to the following question:

If the FDE left tomorrow, would the organization still improve?

If the answer is no, the work is incomplete.

What Has Changed in the Organization

After 12 months, success looks like structural shift, not surface polish.

Reality Is Documented, Not Tribal

- Key workflows are described as they actually operate
- Exceptions are known, not discovered accidentally
- Decisions are traceable
- Past reasoning can be revisited

IRAC+ artifacts exist and are used—not archived.

Ownership Is Clear

- Every deployed system has a named owner
- Escalation paths are explicit
- Responsibility does not route through the FDE
- Maintenance is planned, not reactive

The system no longer depends on a single person's memory.

Automation Is Trusted—but Limited

- Automated steps are narrow and intentional
- Human override paths exist and are exercised
- Failures surface quickly
- Metrics reflect reality, not optimism

The organization understands what the system should not do.

The Organization Learns Faster Than the System Fails

- Failures are surfaced early
- Feedback loops are real, not symbolic
- Iteration happens without drama
- Bad paths are exited deliberately

Progress is measured in reduced surprise, not feature count.

What Has Changed in the Engineer's Role

A successful FDE engagement ends with less dependence, not more.

The FDE Is No Longer the Bottleneck

- Others can explain the system
- Decisions do not require the FDE's presence
- Documentation substitutes for availability
- Knowledge is transferable

This is not a loss of influence.
It is evidence of success.

The FDE Has Shifted from Execution to Oversight (or Exit)

At 12 months, the FDE is either:

- Operating at a higher abstraction level, or
- Preparing to rotate out cleanly

Remaining embedded indefinitely is a warning sign—not an achievement.

What Has Not Happened (And That's the Point)

In a healthy engagement:

- Not everything is automated
- Not every workflow is optimized
- Not every idea was built
- Not every stakeholder is thrilled

What exists is alignment with reality, not illusion of progress.

The Quiet Signal of Success

The strongest signal of success is subtle:

Conversations sound different.

People say:

- "We already know why that happens."
- "That was documented last time."
- "We decided not to automate that—and here's why."
- "We caught this early."

These statements mean the organization now **thinks differently**.

Redefining Professional Pride

For Forward Deployed Engineers, success is not being indispensable.

It is making yourself **safely replaceable**.

If the work requires you to stay forever, something important was not transferred.

Closing Perspective

Forward Deployed Engineering is not about brilliance.

It is about restraint, patience, and fidelity to reality.

After 12 months, the best outcome is not applause.

It is quiet competence that persists after you leave.

Transition to the Closing Note

The final pages of this book are not a summary of techniques.

They are a reminder:

Reality is the hardest system you will ever deploy into—and the only one that matters.

Appendix A Closing Note

From Case to Rule

Each case study tells its own story — Army command posts, corporate adoption battles, hybrid engineers at energy sites, startups trading cash for moats, immigration lines stretching out the door.

Together they teach one rule:
The field is the final exam.
Documents, pitches, and prototypes don't survive without immersion. Adoption isn't won in slides; it's earned in trust. Scaling isn't a graph; it's empathy at work.

Forward Deployed Engineers don't just code. They fix forward, debug under fire, and translate between systems and people.

The cases may change. The environments will differ. But the lesson is constant:

☛ **In the field, theory ends and engineering begins.**

"These cases prove the same rule: the field teaches faster than theory. Whether in government, startups, infrastructure, or global systems, Forward Deployed Engineers win not by writing perfect code, but by showing up where failure hurts."

Appendix B – Templates

"Templates are tools. They save time, but they lie.
Every template must be broken and rebuilt in the field."
— Forward Deployed Engineer Rule Book

Templates and Tools: Shortcuts, Not Solutions

Templates save time. Tools speed work. But neither is the answer.

A template is a map — and every map is wrong. The field will break it. The point is not to follow a template blindly, but to use it as a starting frame, then adapt it until it fits.

- **Rule Template**: "Ship small, ship often." Forged in failure, proven in fire.
- **Scaling with Empathy**: The same tool that works in one office may collapse in five. Tools must scale with people, not just with code.
- **Trust**: A template without adaptation destroys trust. Nothing breaks credibility faster than handing the field a form that doesn't fit reality.

The Rule:

☛ *Templates buy speed. Rule Templates carry scars. Together, they form the FDE's toolkit.*

Core Rule Templates:
- *Ship small, ship often.*
- *Test in the field, not the lab.*
- *Fix forward.*
- *Adoption is the mission.*
- *Scale with empathy.*
- *Immersion before judgment.*

Tools are no different. Every tool is a shortcut. It solves 80% of the problem, and hides the 20% that matters. The FDE's job is not to worship the tool, but to find where it bends, and where it breaks.

Lessons for FDEs:
Use templates to move fast, not to avoid thinking.
Never ship a tool you haven't tested in the field.
A broken template teaches more than a polished one.
Every rule template is written in pain — someone failed so you don't have to.

Why It Matters
An FDE without templates drowns in repetition. An FDE who trusts templates too much drowns in failure. The craft lies in the middle: use the shortcut, then adapt it.

Appendix B – Templates

Templates save time. Tools speed work. But neither is the answer.

A template is a shortcut. A starting frame. A way to avoid reinventing the wheel. But every template is wrong the moment it meets the field.

The FDE uses templates like maps:

- They show the way forward, but the terrain is always different.
- They must be broken, bent, and rebuilt until they fit.

Rule Templates vs. Static Templates

- **Static Templates** are empty forms. They give false comfort.
- **Rule Templates** are forged in pain — scars turned into guidelines.
  - "Ship small, ship often."
  - "Test in the field, not the lab."
  - "Adoption is the mission."

Every rule template carries blood in its ink. Someone failed so you don't have to.

Tools and the 80/20 Trap

Every tool solves 80% of the problem, and hides the 20% that matters.

- Dashboards mask friction until you walk the floor.
- Automation hides workarounds that only humans notice.
- Frameworks collapse when scaled across cultures or sites.

The Rule:

☛ *Use the tool. Find where it bends. Learn where it breaks.*

Lessons for FDEs

1. Templates buy you speed. Adaptation buys you survival.
2. Never ship a tool you haven't tested in the field.
3. A broken template teaches more than a polished one.
4. Rule templates are earned, not invented.

Why It Matters

An FDE without templates drowns in repetition. An FDE who trusts templates too much drowns in failure.

The craft lies in the middle:

☛ **Use the shortcut. Then adapt it.**

Appendix C – Rule Templates

"Rules are scars turned into shortcuts."
— Forward Deployed Engineer Rule Book

Every FDE learns by failing. But failure only becomes progress when you capture it, name it, and reuse it. That's what a **Rule Template** is: a guideline forged in the field, tested in fire, and sharp enough to use again.

Core Rule Templates

- **Ship small, ship often**.

Big-bang releases collapse. Small releases earn trust.

- **Test in the field, not the lab**.

Reality breaks theory. Only the field tells the truth.

- **Fix forward**.

Don't roll back. Patch, improve, adapt — momentum beats perfection.

- **Adoption is the mission**.

Code unused is code that failed. Adoption is the only proof.

- **Scaling with empathy**.

A rollout without training is a rollout doomed. Scale people, not just systems.

- **Immersion before judgment**.

You cannot see the problem from the office. Presence precedes clarity.

How to Use Rule Templates

1. Treat them as starting points, not commandments.
2. Adapt them to the context you're in.
3. Add new rules when failure teaches you something worth saving.
4. Retire rules that no longer serve progress.

The Rule

☛ *A template is a guess. A rule template is a scar. Carry both, but trust the scar first.*

Why It Matters

Rule templates are not theory. They are lessons encoded in shorthand. When you face chaos, they give you a grip. When you face resistance, they give you proof.

An FDE's collection of rule templates is their true toolkit — light enough to carry anywhere, heavy enough to matter everywhere.

Appendix D – Career Roadmap

"The field will shift beneath your feet.
The only constant path
is the one that keeps moving."
— Forward Deployed Engineer Rule Book

The FDE is not a job title. It is a path.

You don't become an FDE by filling out a form. You become one by walking the field, failing forward, and turning scars into shortcuts. This roadmap is not exhaustive, but it is tested.

Stage 1 – Student: Building the Hidden Toolkit

- **Visible Degree**: Industry-specific credential (CPA, Real Estate License, CDL, FAA, insurance producer license).
- **Secret Degree**: Computer Science / Software Engineering.
- **Mindset**: Apprentice. You are preparing for immersion.
- **Goal**: Earn the keys that let you enter the system.

Stage 2 – Off the Street Hire: Insider Immersion

- **Entry Path**: You join as 'just another employee.' **Caution**: You may not land in your target unit. Transfers are slow, opportunities scarce, mobility uncertain. Your secret degree gives you agency when placement is out of your control.
- **Skin in the Game**: Your paycheck, license, and reputation are on the line.
- **Risk**: Assimilation. Forgetting why you came.
- **Goal**: Map the system from the inside. Learn its culture, politics, and pain points.

Stage 3 – Field Apprentice: Practicing Presence

- Immersion: Shadow the workflows. Read the field, not the reports.
- Fix Forward: Patch what you can, even if imperfect.
- Lessons: Every broken process is a teacher. Every workaround is a signal.
- Goal: Develop instincts only immersion can give.

Stage 4 – Recognized FDE: Translating & Delivering

- Bridge Role: You are the connector between field pain and engineering solution.
- Debugging Under Fire: Fixes are no longer theoretical — they must work live.
- Adoption Accelerator: You make change stick by being present when it fails.
- Goal: Deliver outcomes, not slides. Build trust, not decks.

Stage 5 – Startup Builder: Trading Margin for Moat

- Moat vs. Margin: You burn resources to embed, knowing presence is the moat.
- Rule Templates: You forge scars into rules for your team.
- Scaling with Empathy: You expand without breaking people.
- Goal: Build systems that outlast you.

Stage 6 – Leader & Teacher: Passing the Compass

- Apprenticeship Never Ends: You are still learning, but now others learn from you.
- Rule Maker / Rule Breaker: You know when to stabilize and when to disrupt.

- Trust: Your reputation is your currency.
- Goal: Multiply FDEs. Leave the field stronger than you found it.

The Rule of the Roadmap

☛ *The FDE path is not linear. You will circle back. Failures force restarts. Success pushes you forward. The only constant is this: the field is your teacher, and you are always its apprentice.*

■ Career Roadmap – Quick Guide

Stage 1 – Student

Earn the *visible credential* (CPA, CDL, license) and carry the *secret degree* (CS/SE) as your hidden lens.

Stage 2 – Off the Street Hire

Enter as "just another employee," but with **skin in the game** — paycheck, license, and reputation.

Stage 3 – Field Apprentice

Shadow the workflows. Fail, fix forward, and let immersion teach what no report can.

Stage 4 – Recognized FDE

Bridge the gap: translate problems into products, and deliver fixes live where failure hurts.

Stage 5 – Startup Builder

Burn margin to build moats. Scale with empathy. Forge scars into **rule templates** for others.

Stage 6 – Leader & Teacher

Multiply yourself. Pass the compass. Leave the field stronger than you found it.

Case File #4 — Recovery Without Rewriting History

Context

The system failed publicly.
Not as a bug, but as an outcome.

Users were affected. Leadership was questioned. External scrutiny followed. Internal confidence dropped. The organization's first instinct was not to fix the system, but to explain it — to soften language, reorder timelines, and identify a narrative that reduced embarrassment.

The Forward Deployed Engineer was brought in after the failure, not before it.

The Trap

When failure is visible, organizations confuse recovery with reputation management.

They attempt to:

- Reframe decisions as misunderstandings
- Reclassify warnings as unforeseeable
- Attribute outcomes to edge cases or misuse
- Replace records with summaries

None of these repair the system.
They only make future learning impossible.

The Bad Move

Leadership requested a rewritten post-mortem.

The original documentation:

- Showed uncertainty at the time of decision
- Recorded unresolved risks
- Captured dissenting views
- Preserved incomplete information

The request was simple and subtle:

"Clean this up so it reflects what we now know."

That request would have erased the only reliable evidence of how the system actually failed.

The Consequence

Rewriting history would have produced:

- A comforting narrative
- Apparent decisiveness
- Short-term reputational relief

It would also have destroyed:

- Trust with operators
- Accountability for decision-making
- The ability to distinguish error from negligence
- Any chance of preventing recurrence

The organization would have recovered its story — and lost its system.

The Fix

The FDE refused to rewrite.

Instead:

- The original record was preserved intact
- A separate recovery document was created
- All corrections were clearly timestamped and labeled as post-failure learning
- Ownership was reassigned without retroactive justification
- The system was repaired without altering the decision trail

No statements were softened.
No intent was inferred after the fact.
No ambiguity was resolved retroactively.

Rule
Recovery happens forward.
History remains intact.

If learning requires erasing the past, it is not learning — it is liability management.

What This Protects

- Institutional memory
- Operator trust
- Future decision-makers
- The distinction between error and misconduct

Most importantly, it preserves the organization's ability to tell the truth to itself under pressure.

Why This Is an FDE Responsibility
At this stage of the career, the FDE is no longer optimizing outcomes.
They are protecting the system's capacity to learn.

That means accepting discomfort.
Allowing scars to remain visible.
And choosing credibility over reassurance.

Closing
Recovery does not mean controlling the story.
It means refusing to corrupt it.

The Forward Deployed Engineer leaves the record intact, repairs the system in daylight, and moves forward without laundering history. Trust is rebuilt not by reassurance, but by continuity — showing that learning survives embarrassment, and accountability survives pressure.

When the system stabilizes, the work is complete.
The scar remains.

That is the point.

Appendix E – Reading & Training

"An FDE's apprenticeship never ends.
Learning is not a phase.
It is the work itself."
— Forward Deployed Engineer Rule Book

The field changes. Tools change. Problems change. Only learning keeps pace. This appendix is not a syllabus — it is a compass. Use it to find what sharpens you.

Core Readings (Mindset)

- **Peter Drucker – The Effective Executive** → Why effectiveness matters more than effort.
- **Andy Grove – High Output Management** → Lessons in scaling people and systems.
- **Ben Horowitz – The Hard Thing About Hard Things** → Building and surviving under pressure.
- **John Boyd – A Discourse on Winning and Losing (OODA Loops)** → Adapting faster than your environment.
- **Nassim Taleb** – Skin in the Game → Why credibility requires personal risk.

Core Readings (FDE History & Practice)

- **Palantir case studies** → The origin of the role. Immersion + mandate.
- **GitLab handbook** → Transparency and adoption as the mission.
- **Carbon/A16Z essays** → Trading margin for moat.
- **Government reports on immigration** → Systems under stress at scale.

Topics:
- Lean prototyping, MVPs, test-and-learn cycles
- Safe experimentation frameworks
- Iterative reporting and field feedback

Exercise: 48-hour sprint: prototype → deploy → collect feedback → present lessons learned.

Phase VI – Communication & Stakeholder Management

Objective: Translate technical insight into executive action.

Topics:
- Writing field reports and executive briefs
- Negotiation, persuasion, and counter-proposal facilitation
- Public vs private discourse strategies

Exercise: Present a mock "Findings & Options" briefing to a simulated board.

Evaluation: Clarity, diplomacy, and actionability of recommendations.

Phase VII – Ethics, Continuity & Custody

Objective: Operate safely in environments with sensitive or classified information.

Topics:
- Chain-of-custody for field notes
- Legal duties to report violations
- Continuity planning and handoff documentation

Exercise: Draft a continuity package for a departing FDE, ensuring operational safety and compliance.

Phase VIII – Capstone Deployment

Objective: Apply every principle in a live environment.

Format: 6–8 week mentored engagement with a real or simulated client.

Deliverables:
- Field Notebook (diagnostics, design decisions, iterations)
- Deployment Log (technical + organizational metrics)
- Post-mortem and debrief presentation

Evaluation: 360-degree review by mentor, peer, and client.

Phase IX – Reflection & Mentorship

Objective: Transition from practitioner to teacher.

Topics:
- Codifying lessons learned into new rules
- Mentoring junior FDEs
- Institutional knowledge capture

Exercise: Write one new "Rule Template" derived from your deployment and defend it before peers.

Certification Path
- Associate FDE: Completion of Phases I–IV
- Certified FDE: Completion of Phases V–VIII
- Senior FDE / Instructor: Completion of Phase IX + two verified deployments

Program Governance
- Continuous improvement via after-action reviews
- Rule Book updates each training cycle
- Repository of anonymized field notebooks for study

■ APPENDIX F

Training Blueprint: How Forward Deployed Engineers Are Actually Formed

Forward Deployed Engineers are not created through certification, bootcamps, or short training programs.

They are formed through exposure, responsibility, and time in the field.
This appendix defines the training blueprint required to produce engineers capable of operating inside real systems—where failure has consequences and abstraction breaks down.

This is not aspirational.

It is descriptive of what actually works.

F.1 **Entry Paths and Prerequisites**

Forward Deployed Engineers begin with dual credentials.

They must hold:

- A strong technical foundation (software, systems, data, or AI)
- A legitimate, field-recognized credential in the domain they serve

Examples include:

- JD + Software Engineering
- MD + Data Science
- CPA or CFA + Computer Science
- Licensed tradesperson + Systems or Automation engineering

This dual identity is not optional.

It is the price of trust.

Without it, the engineer is treated as an outsider.

With it, they are allowed to observe real work, hear unfiltered stories, and see how systems actually behave under pressure.

Entry may occur in one of three ways:

- Hired inside and trained slowly
- Invited in as an external specialist
- Hybrid, evolving from one to the other

Each path trades speed for depth.

None bypass immersion.

F.2 **Apprenticeship Time Horizons**

Complex systems cannot be understood quickly.

Organizations often underestimate how long it takes to learn:

- Informal workflows
- Exception handling
- Tacit rules
- Political constraints
- Failure recovery behaviors

Forward Deployed Engineers therefore train as apprentices, whether or not the organization names it as such.

Apprenticeship includes:

- Extended observation before intervention
- Repetition across many cases
- Exposure to edge conditions
- Gradual increase in responsibility
- Accountability for outcomes, not just effort

Time spent here is not inefficiency.

It is **risk reduction**.

Engineers who skip this phase produce brittle systems that fail quietly and expensively.

F.3 **Field Training Pillars**

Training a Forward Deployed Engineer is not classroom instruction.
It is **field conditioning.**

Effective training rests on four pillars:

1. **Observation Before Action**

The engineer must learn to watch without fixing.

This includes:

- Shadowing frontline staff
- Witnessing breakdowns in real time
- Observing workarounds instead of dismissing them
- Learning when not to intervene

Premature optimization destroys signal.

2. **Elicitation Without Authority**

Forward Deployed Engineers are trained to extract accurate information by observing, listening, and validating.

They learn:

- How memory distorts recall
- Why certainty is unreliable
- How leading questions poison requirements
- Why contradictions are valuable signals

Truth emerges gradually and indirectly.

Training emphasizes patience, neutrality, and disciplined questioning.

3. **Documentation as Custody of Insight**

Engineers are trained to treat documentation as **continuity infrastructure**, not bureaucracy.

Notes must preserve:

- Context
- Reasoning
- Uncertainty
- Tradeoffs

This prevents insight loss when:

- Engineers rotate out
- Projects pause
- Systems change hands

Training emphasizes recording thinking, not just conclusions.

4. **Fix-Forward Discipline**

Trainees learn to:

- Make bounded changes
- Ship incremental improvements
- Observe impact in real conditions
- Iterate without rollback panic

Perfection is not the goal.
Durability is.

F.4 **Documentation and Continuity Artifacts**

Forward Deployed Engineers are trained to leave behind transferable artifacts, not just working code.

These artifacts:

- Allow another engineer to resume work without re-immersion
- Make decisions auditable
- Preserve institutional memory
- Reduce dependency on individual heroics

Documentation standards are treated as part of training—not optional hygiene.

An engineer who cannot leave continuity behind is not yet fully trained.

F.5 Evaluation and Readiness Signals

Forward Deployed Engineers are not evaluated by speed alone.

Readiness signals include:

- Ability to operate independently without escalating prematurely
- Comfort admitting uncertainty
- Discipline in documentation
- Trust earned from operators, not just managers
- Judicious use of authority
- Willingness to delay automation when learning is incomplete

Early autonomy is a warning sign, not a success metric.

Engineers are considered field-ready when they can:

- Diagnose without blaming
- Intervene without destabilizing
- Exit without leaving damage behind

F.6 **What This Blueprint Is (and Is Not)**

This blueprint is not:

- A certification program
- A bootcamp
- A checklist
- A shortcut

It is:

- A realistic description of how capable field engineers are formed
- A warning against premature deployment
- A framework for protecting both the engineer and the organization

Forward Deployed Engineering is a craft.

It cannot be rushed.
It cannot be simulated.
And it cannot be learned from a distance.

Closing Note

Organizations that invest in this kind of training do not just build better systems.

They build **institutional memory**, **operational resilience**, and **trust that survives failure**.

That is the real return on Forward Deployed Engineering.

APPENDIX G — INTERVIEWING & ELICITATION REFERENCE

Field Guide for Accurate Information Extraction

Purpose of This Appendix

This appendix exists to support—not replace—the main text.

It is a reference, not a script.

The Five Elicitation Principles

1. Accuracy beats speed
2. Narratives precede structure
3. Artifacts outrank memory
4. Contradictions are signals
5. Documentation preserves insight

Question Types (When to Use Each)

Open-Ended

- "Walk me through..."
- "Describe the last time..."

Use for: surfacing mental models

Clarifying

- "What happened next?"
- "Who handles that?"

Use for: sequencing and precision

Probing

- "How often does that occur?"
- "What happens when this fails?"

Use for: uncovering edge cases

Verification

- "Can you show me?"
- "Where is that recorded?"

Use for: artifact alignment

Prohibited (Do Not Use Early)

- Leading questions
- Compound questions
- Yes/no framing
- Hypotheticals
- Solution proposals

Memory Pitfalls to Expect

- Confidence without accuracy
- Timeline compression
- Post-hoc rationalization
- Policy–practice mismatch
- Exception blindness

Interview Red Flags

- Perfectly clean explanations
- No remembered failures
- "That never happens"
- Excessive abstraction
- Defensiveness when asked for artifacts

Post-Interview Checklist

- IRAC+ drafted same day
- Context fully captured
- Artifacts referenced
- Uncertainties noted
- Next elicitation identified

APPENDIX H — IRAC+ TEMPLATES

Continuity-Safe Documentation for FDEs

IRAC+ MASTER TEMPLATE

Context

- Person(s):
- Place/System:
- Date/time
- Situation:
- Constraints:
- Artifacts:

Issue

- What is the problem or risk?
- Why does it matter now?

Rule

- Which FDE Rule(s) apply?
- Which external constraints apply?

Application

- How the rule interacts with reality
- Conflicts, tradeoffs, uncertainty

Conclusion

- Proposed fix / experiment / decision
- Scope and next step

Single-Issue Example (Structure Only)

- One context
- One issue
- One rule
- One bounded conclusion

Multi-Issue Example (Structure Only)

- Shared context
- Multiple issues (clearly separated)
- Rules applied per issue
- Independent conclusions

Handoff-Ready Checklist

- Can another engineer re-run this investigation?
- Are assumptions explicit?
- Is access documented?
- Are conclusions revisitable?
- Are artifacts traceable?

If not, the IRAC+ is incomplete.

Reader Roadmap

Rule Index: Field Principles of the Forward Deployed Engineer

#	Rule Template / Principle	Summary
1	Ask the Five Ws Before You Code	Never diagnose from a dashboard; walk the floor and let the field speak.
2	Fix Forward	Act quickly with reversible decisions; learn faster than bureaucracy can react.
3	Authority Follows Competence	The right to act is earned by understanding the system better than anyone else.
4	Build a Moat Before You Leave	Each deployment must leave behind a defensible, self-sustaining capability.
5	Design for Continuity	Document so another FDE can resume without restarting the mission.
6	Speak Both Languages	Translate between technical and executive dialects with equal fluency.
7	Expose Problems to Daylight	Encourage open discourse with decision-makers; avoid hidden fixes.
8	Prototype Under Fire	Deliver small, live experiments instead of perfect static plans.
9	Report What You Find — No Exceptions	Ethical reporting is part of deployment; silence transfers liability.
10	Train the Next Deployed	Every mission ends with a successor brief—no field without follow-through.

✦ Closing Note ✦

"The Forward Deployed Engineer does not wait for permission.
They go where the problems live.
They listen before they fix.
They build with the field, not above it.
And when the work is done,
they move forward — ready for the next unknown."
— Forward Deployed Engineer Rule Book

The FDE is not a title. It is a posture: step into the system, shoulder the load, fix forward.

The work will never be easy.
It will rarely be clean.
But it will always matter.

Carry the rules. Test them. Break them. Forge new ones — not as doctrine, but as tools. Shape them, bend them, test them in the fire of real deployments. What survives is yours.

And remember: in every field, in every challenge, in every unknown — the rule is the same.

Go where the work is.
Learn what the field teaches.
Fix forward.

📜 The FDE Rule Poster

Forward Deployed Engineer (FDE) – Rules at a Glance

Mandate & Authority

1. **Appointed, not hired** – Mandated by CEO or Congress.
2. **Report to the top** – Direct line to leadership.
3. **Absolute access** – No silos, full system view.
4. **Neutral position** – Mission > politics.

Training & Rotation

1. **Dual credentials** – Engineering + domain expertise.
2. **Full immersion** – Complete insider training.
3. **Rotations required** – Every branch, every angle.
4. **Learning log** – Capture notes, maps, prototypes.

Prototyping & Testing

1. **Clarity before code** – Specify → Plan → Task → Implement.
2. **One problem, one prototype** – Keep it small, testable.
3. **Simulate first** – Sandbox before real-world use.
4. **Validation gates** – Don't move forward until tested.

Outcomes & Measurement

1. **Outcome, not output** – Measure problems solved.
2. **Contracts grow with valu**e – Bigger impact = bigger mandate.
3. **Cost/value must improve** – Solve more for less.
4. **Generalize lessons** – Make fixes reusable.

Culture & Integrity

1. **Pain is part of the job** – Expect resistance.
2. **Stay mission-first** – Never settle for "safe."
3. **Resist consulting drift** – Solve real problems, not surface requests.
4. **Publish findings** – Deliver the full system map.

🏗 FDE Motto

"Do things that don't scale — at scale."

Made in the USA
Las Vegas, NV
13 June 2026

956a3abe-e1d5-42bd-b7d4-0d0d770a2790R01